Data Analysis

Data Analysis

A Bayesian Tutorial

Second Edition

D.S.Sivia

Rutherford Appleton Laboratory and St. Catherine's College, Oxford

with

J.Skilling

Maximum Entropy Data Consultants, Cambridge

Data Analysis

A Bayesian Tutorial

D.S.Sivia

Rutherford Appleton Laboratory and St. Catherine's College, Oxford

with

J.Skilling

Maximum Entropy Data Consultants, Cambridge

OXFORD
UNIVERSITY PRESS

OXFORD

UNIVERSITY PRESS

Great Clarendon Street, Oxford OX2 6DP
United Kingdom

Oxford University Press is a department of the University of Oxford.
It furthers the University's objective of excellence in research, scholarship,
and education by publishing worldwide. Oxford is a registered trade mark of
Oxford University Press in the UK and in certain other countries

First published 2006
Reprinted 2012

British Library Cataloguing in Publication Data
Data available

Library of Congress Cataloging in Publication Data
Data available

ISBN 978-0-19-856832-2

Printed and bound by CPI Group (UK) Ltd, Croydon, CR0 4YY

To Joyce and Jennifer

Preface

As an undergraduate, I always found the subject of statistics to be rather mysterious. This topic wasn't entirely new to me, as we had been taught a little bit about probability earlier at high school; for example, I was already familiar with the binomial, Poisson and normal distributions. Most of this made sense, but only seemed to relate to things like rolling dice, flipping coins, shuffling cards and so on. However, having aspirations of becoming a scientist, what I really wanted to know was how to analyse experimental data. Thus, I eagerly looked forward to the lectures on statistics. Sadly, they were a great disappointment. Although many of the tests and procedures expounded were intuitively reasonable, there was something deeply unsatisfactory about the whole affair: there didn't seem to be any underlying basic principles! Hence, the course on 'probability and statistics' had led to an unfortunate dichotomy: probability made sense, but was just a game; statistics was important, but it was a bewildering collection of tests with little obvious rhyme or reason. While not happy with this situation, I decided to put aside the subject and concentrate on real science. After all, the predicament was just a reflection of my own inadequacies and I'd just have to work at it when the time came to really analyse my data.

The story above is not just my own, but is the all too common experience of many scientists. Fortunately, it doesn't have to be like this. What we were not told in our undergraduate lectures is that there is an alternative approach to the whole subject of data analysis which uses only probability theory. In one sense, it makes the topic of statistics entirely superfluous. In another, it provides the logical justification for many of the prevalent statistical tests and procedures, making explicit the conditions and approximations implicitly assumed in their use.

This book is intended to be a tutorial guide to this alternative Bayesian approach, including modern developments such as maximum entropy. It is not designed to be a comprehensive or authoritative text, but is aimed at showing the reader how simple and straightforward it is to use probability theory directly for data analysis. As such, the examples are to be taken as illustrative rather than exhaustive; to attempt the latter would encourage the cook-book recipe mentality which bedevils conventional statistics. It is hoped that the emphasis on going from a few basic principles to a concrete statistical prescription will provide the reader with the confidence to tackle problems which are not found in text books. If he, or she, comes away with the feeling: 'Is that all it is? It's easy — I can do that', then the enterprise can be deemed to have been successful.

This book should be accessible to all students who have completed a standard first year undergraduate course in Mathematical Methods, covering topics such as partial derivatives, multiple integrals, matrices, vectors and so on; this should certainly encompass most physicists, chemists and engineers, and quite a few others besides. The first three chapters form the core material, and deal with the basic principles and the subject of parameter estimation; in fact, most data analysis problems require no more than this. Chapter 4 provides a valuable insight into hypothesis testing and Chapter 5 gives an alternative rationale for many familiar probability distributions. Chapter 6, which looks at non-parametric estimation, is perhaps the most demanding and is also liable to change significantly with future developments; along with Chapter 7 on experimental design, it is really for the enthusiast.

Finally, it is my pleasure to acknowledge the invaluable suggestions and comments given to me by several friends and colleagues: in particular, Phil Gregory, Steve Gull, Kevin Knight, Sabine König, Angus Lawson, Jerry Mayers, Vincent Macaulay, Toby Perring, Roger Pynn, Steve Rawlings, John Skilling and Dave Waymont. While I may yet regret not taking all their advice, this text has been improved greatly because of them; I thank them for their patience and help.

Rutherford Appleton Laboratory
October, 1995   D.S.S.

Preface to the Second Edition

A couple of years ago, I was approached by some colleagues about the possibility of updating my book. My initial reaction was to think why should I? For what I had intended, I would still write it pretty much the same way. Upon reflection, however, the merits of a Second Edition became apparent. For example, it would allow me to include the simple ideas for robust estimation that I had explored since the original text; namely, ways of dealing with 'outliers', and unknown correlated noise, within the framework of the ubiquitous least-squares procedure. More importantly, though, John Skilling had kindly offered to contribute a couple of chapters on modern numerical techniques for Bayesian computation. The latter have taken longer to materialize than anticipated, because John's thoughts on the topic have been continually evolving over the period. Indeed, the subject is at the forefront of his current research.

The strengths of the original book have been preserved by keeping the existing text largely unaltered, but it has been enhanced with the addition of Chapters 8, 9 and 10. Even with the extra material, including a new Appendix, the aim has been to keep the volume slim. This has been achieved through the type-setting efficiency of LaTeX, and some culling of content now deemed less useful.

It goes without saying that I am greatly indebted to John Skilling, my co-author, for Chapters 9 and 10, Appendix B, and for suggestions on changes to the original text. John has included examples of 'C' computer code in his chapters to illustrate the main ideas. These short programs are freely available in the public domain, for use without warranty, under the terms of the *GNU General Public License*; for details see http://www.gnu.org/copyleft/gpl.html. Courtesy suggests, of course, that their use be acknowledged by reference to this volume (*D. S. Sivia with J. Skilling, 'Data Analysis — a Bayesian tutorial', 2nd Edition, 2006, Oxford University Press*). Related material should be available at http://www.inference.phy.cam.ac.uk/bayesys/, and I am very grateful to David MacKay for this use of his website.

Finally, I'd like to acknowledge the valuable feedback I was given on some of the new material by Richard Bailey, Bruce Henning, Steve Rawlings, Stephen Stokes and David Waymont, and the help and patience of my Editor, Sönke Adlung, at OUP.

Oxford
October, 2005   D.S.S.

Contents

Part I

The essentials

'La théorie des probabilités n'est que le bon sens reduit au calcul.'

— Pierre Simon, Marquis de Laplace

1 The basics

'There are three kinds of lies: lies, damned lies and statistics.'

Mark Twain (1924) probably had politicians in mind when he reiterated Disraeli's famous remark. Scientists, we hope, would never use data in such a selective manner to suit their own ends. But, alas, the analysis of data is often the source of some exasperation even in an academic context. On hearing comments like 'the result of this experiment was inconclusive, so we had to use statistics', we are frequently left wondering as to what strange tricks have been played on the data.

The sense of unease which many of us have towards the subject of statistics is largely a reflection of the inadequacies of the 'cook-book' approach to data analysis that we are taught as undergraduates. Rather than being offered a few clear principles, we are usually presented with a maze of tests and procedures; while most seem to be intuitively reasonable individually, their interrelations are not obvious. This apparent lack of a coherent rationale leads to considerable apprehension because we have little feeling for which test to use or, more importantly, why.

Fortunately, data analysis does not have to be like this! A more unified and logical approach to the whole subject is provided by the probability formulations of Bayes and Laplace. Bayes' ideas (published in 1763) were used very successfully by Laplace (1812), but were then allegedly discredited and largely forgotten until they were rediscovered by Jeffreys (1939). In more recent times they have been expounded by Jaynes (1983, 2003) and others. This book is intended to be an introductory tutorial to the Bayesian approach, including modern developments such as maximum entropy.

1.1 Introduction: deductive logic versus plausible reasoning

Let us begin by trying to get a general feel for the nature of the problem. A schematic representation of deductive logic is shown in Fig. 1.1(a): given a cause, we can work out its consequences. The sort of reasoning used in pure mathematics is of this type: that is to say, we can derive many complicated and useful results as the logical consequence of a few well-defined axioms. Everyday games of chance also fall into this category. For example, if we are told that a fair coin is to be flipped ten times, we can calculate the chances that all ten tosses will produce heads, or that there will be nine heads and one tail, and so on.

Most scientists, however, face the reverse of the above situation: Given that certain effects have been observed, what is (are) the underlying cause(s)? To take a simple example, suppose that ten flips of a coin yielded seven heads: Is it a fair coin or a biased one? This type of question has to do with inductive logic, or plausible reasoning, and is illustrated schematically in Fig. 1.1(b); the greater complexity of this diagram is

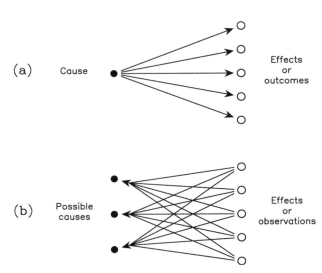

Fig. 1.1 A schematic representation of (a) deductive logic, or pure mathematics, and (b) plausible reasoning, or inductive logic.

designed to indicate that it is a much harder problem. The most we can hope to do is to make the best inference based on the experimental data and any prior knowledge that we have available, reserving the right to revise our position if new information comes to light. Around 500 BC, Herodotus said much the same thing: 'A decision was wise, even though it led to disastrous consequences, if the evidence at hand indicated it was the best one to make; and a decision was foolish, even though it led to the happiest possible consequences, if it was unreasonable to expect those consequences.'

Even though plausible reasoning is rather open-ended, are there any general quantitative rules which apply for such inductive logic? After all, this issue is central to data analysis.

1.2 Probability: Cox and the rules for consistent reasoning

In 1946, Richard Cox pondered the quantitative rules necessary for logical and consistent reasoning. He started by considering how we might express our relative beliefs in the truth of various propositions. For example: (a) it will rain tomorrow; (b) King Harold died by being hit in the eye with an arrow at the battle of Hastings in 1066 AD; (c) this is a fair coin; (d) this coin is twice as likely to come up heads as tails; and so on. The minimum requirement for expressing our relative beliefs in the truth of these propositions in a consistent fashion is that we rank them in a *transitive* manner. In other words, if we believe (a) more than (b), and (b) more than (c), then we must necessarily believe (a) more than (c); if this were not so, we would continue to argue in circles. Such a transitive ranking can easily be obtained by assigning a *real* number to each of the propositions in a manner so that the larger the numerical value associated with a proposition, the more we believe it.

Cox actually took this much for granted — as being obvious — and wondered what rules these numbers had to obey in order to satisfy some simple requirements of logical consistency. He began by making two assertions. The first is very straightforward: if we specify how much we believe that something is true, then we must have implicitly specified how much we believe it's false. He didn't assume any particular form for this relationship, but took it as being reasonable that one existed. The second assertion is slightly more complicated: if we first specify how much we believe that (proposition) Y is true, and then state how much we believe that X is true given that Y is true, then we must implicitly have specified how much we believe that both X and Y are true. Again, he only asserted that these quantities were related but did not specify how. To work out the actual form of the relationships, Cox used the rules of *Boolean logic*, ordinary algebra, and the constraint that if there were several different ways of using the same information then we should always arrive at the same conclusions irrespective of the particular analysis-path chosen. He found that this consistency could only be ensured if the real numbers we had attached to our beliefs in the various propositions obeyed the usual rules of probability theory:

$$\mathrm{prob}(X|I) + \mathrm{prob}(\overline{X}|I) = 1 \qquad (1.1)$$

and

$$\mathrm{prob}(X, Y|I) = \mathrm{prob}(X|Y, I) \times \mathrm{prob}(Y|I), \qquad (1.2)$$

with $0 = \mathrm{prob}(\mathrm{false})$ and $1 = \mathrm{prob}(\mathrm{true})$ defining certainty; a proof is given in Appendix B. Here $\overline{X}$ denotes the proposition that X is false, the vertical bar ' $|$ ' means 'given' (so that all items to the right of this conditioning symbol are taken as being true) and the comma is read as the conjunction 'and'.

We have made the probabilities conditional on I, to denote the relevant background information at hand, because there is no such thing as an absolute probability. For example, the probability we assign to the proposition 'it will rain this afternoon' will depend on whether there are dark clouds or a clear blue sky in the morning; it will also be affected by whether or not we saw the weather forecast. Although the conditioning on I is often omitted in calculations, to reduce algebraic cluttering, we must never forget its existence. A failure to state explicitly all the relevant background information, and assumptions, is frequently the real cause of heated debates about data analysis.

Equation (1.1) is called the *sum rule*, and states that the probability that X is true plus the probability that X is false is equal to one. Equation (1.2) is called the *product rule*. It states that the probability that both X and Y are true is equal to the probability that X is true given that Y is true times the probability that Y is true (irrespective of X). We can change the description of probability, as when we multiply by 100 to obtain percentages, but we are not allowed to change the content of eqns (1.1) and (1.2). Probability calculus uses the unique scale in which the rules take the form of un-adorned addition and multiplication.

1.3 Corollaries: Bayes' theorem and marginalization

The sum and product rules of eqns (1.1) and (1.2) form the basic algebra of probability theory. Many other results can be derived from them. Amongst the most useful are two known as *Bayes' theorem* and *marginalization*:

$$\text{prob}(X|Y,I) = \frac{\text{prob}(Y|X,I) \times \text{prob}(X|I)}{\text{prob}(Y|I)} \qquad (1.3)$$

and

$$\text{prob}(X|I) = \int_{-\infty}^{+\infty} \text{prob}(X,Y|I)\,\mathrm{d}Y. \qquad (1.4)$$

Bayes' theorem, or eqn (1.3), follows directly from the product rule. To see this, let's rewrite eqn (1.2) with X and Y *transposed* (or interchanged):

$$\text{prob}(Y,X|I) = \text{prob}(Y|X,I) \times \text{prob}(X|I).$$

Since the statement that 'Y and X are both true' is the same as 'X and Y are both true', so that $\text{prob}(Y,X|I) = \text{prob}(X,Y|I)$, the right-hand side of the above can be equated to that of eqn (1.2); hence, we obtain eqn (1.3). It is invaluable because it enables us to turn things around with respect to the conditioning symbol: it relates $\text{prob}(X|Y,I)$ to $\text{prob}(Y|X,I)$. The importance of this property to data analysis becomes apparent if we replace X and Y by *hypothesis* and *data*:

$$\text{prob}(hypothesis|data,I) \propto \text{prob}(data|hypothesis,I) \times \text{prob}(hypothesis|I).$$

The power of Bayes' theorem lies in the fact that it relates the quantity of interest, the probability that the hypothesis is true given the data, to the term we have a better chance of being able to assign, the probability that we would have observed the measured data if the hypothesis was true.

The various terms in Bayes' theorem have formal names. The quantity on the far right, $\text{prob}(hypothesis|I)$, is called the *prior* probability; it represents our state of knowledge (or ignorance) about the truth of the hypothesis before we have analysed the current data. This is modified by the experimental measurements through the *likelihood* function, or $\text{prob}(data|hypothesis,I)$, and yields the *posterior* probability, $\text{prob}(hypothesis|data,I)$, representing our state of knowledge about the truth of the hypothesis in the light of the data. In a sense, Bayes' theorem encapsulates the process of learning. We should note, however, that the equality of eqn (1.3) has been replaced with a proportionality, because the term $\text{prob}(data|I)$ has been omitted. This is fine for many data analysis problems, such as those involving *parameter estimation*, since the missing denominator is simply a normalization constant (not depending explicitly on the hypothesis). In some situations, like *model selection*, this term plays a crucial rôle. For that reason, it is sometimes given the special name of *evidence*. This crisp single word captures the significance of the entity, as opposed to older names, such as *prior predictive* and *marginal likelihood*, which describe how it tends to be used or calculated. Such a central quantity ought to have a simple name, and 'evidence' has been assigned no other technical meaning (apart from as a colloquial synonym of data).

The marginalization equation, (1.4), should seem a bit peculiar: up to now Y has stood for a given proposition, so how can we integrate over it? Before we answer that question, let us first consider the marginalization equation for our standard X and Y propositions. It would take the form

$$\text{prob}(X|I) = \text{prob}(X,Y|I) + \text{prob}(X,\overline{Y}|I). \qquad (1.5)$$

This can be derived by expanding $\text{prob}(X,Y|I)$ with the product rule of eqn (1.2):

$$\text{prob}(X,Y|I) = \text{prob}(Y,X|I) = \text{prob}(Y|X,I) \times \text{prob}(X|I),$$

and adding a similar expression for $\text{prob}(X,\overline{Y}|I)$ to the left- and right-hand sides, respectively, to give

$$\text{prob}(X,Y|I) + \text{prob}(X,\overline{Y}|I) = \left[\text{prob}(Y|X,I) + \text{prob}(\overline{Y}|X,I)\right] \times \text{prob}(X|I).$$

Since eqn (1.1) ensures that the quantity in square brackets on the right is equal to unity, we obtain eqn (1.5). Stated verbally, eqn (1.5) says that the probability that X is true, irrespective of whether or not Y is true, is equal to the sum of the probability that both X and Y are true and the probability that X is true and Y is false.

Suppose that instead of having a proposition Y, and its negative counterpart $\overline{Y}$, we have a whole set of alternative possibilities: $Y_1, Y_2, \ldots, Y_M = \{Y_k\}$. For example, let's imagine that there are M (say five) candidates in a presidential election; then Y_1 could be the proposition that the first candidate will win, Y_2 the proposition that the second candidate will win, and so on. The probability that X is true, for example that unemployment will be lower in a year's time, irrespective of whoever becomes president, is then given by

$$\text{prob}(X|I) = \sum_{k=1}^{M} \text{prob}(X,Y_k|I). \qquad (1.6)$$

This is just a generalization of eqn (1.5), and can be derived in an analogous manner, by putting $\text{prob}(X,Y_k|I) = \text{prob}(Y_k|X,I) \times \text{prob}(X|I)$ in the right-hand side of eqn (1.6), as long as

$$\sum_{k=1}^{M} \text{prob}(Y_k|X,I) = 1. \qquad (1.7)$$

This *normalization* requirement is satisfied if the $\{Y_k\}$ form a *mutually exclusive* and *exhaustive* set of possibilities. That is to say, if one of the Y_k's is true then all the others must be false, but one of them has to be true.

The actual form of the marginalization equation in (1.4) applies when we go to the *continuum limit*. For example, when we consider an arbitrarily large number of propositions about the range in which (say) the Hubble constant H_0 might lie. As long as we choose the intervals in a contiguous fashion, and cover a big enough range of values for H_0, we will have a mutually exclusive and exhaustive set of possibilities. Equation (1.4) is then just a generalization of eqn (1.6), with $M \to \infty$, where we have used the usual shorthand notation of *calculus*. In this context, Y now represents the numerical value of a parameter of interest (such as H_0) and the integrand $\text{prob}(X,Y|I)$

is technically a *probability density* function rather than a probability. Strictly speaking, therefore, we should denote it by a different symbol, such as $\text{pdf}(X, Y | I)$, where

$$\text{pdf}(X, Y = y | I) = \lim_{\delta y \to 0} \frac{\text{prob}(X, y \leqslant Y < y + \delta y | I)}{\delta y}, \qquad (1.8)$$

and the probability that the value of Y lies in a finite range between y_1 and y_2 (and X is also true) is given by

$$\text{prob}(X, y_1 \leqslant Y < y_2 | I) = \int_{y_1}^{y_2} \text{pdf}(X, Y | I) \, \mathrm{d}Y. \qquad (1.9)$$

Since 'pdf' is also a common abbreviation for *probability distribution* function, which can pertain to a discrete set of possibilities, we will simply use 'prob' for anything related to probabilities; this has the advantage of preserving a uniformity of notation between the continuous and discrete cases. Thus, in the continuum limit, the normalization condition of eqn (1.7) takes the form

$$\int_{-\infty}^{+\infty} \text{prob}(Y | X, I) \, \mathrm{d}Y = 1. \qquad (1.10)$$

Marginalization is a very powerful device in data analysis because it enables us to deal with *nuisance parameters*; that is, quantities which necessarily enter the analysis but are of no intrinsic interest. The unwanted background signal present in many experimental measurements, and instrumental parameters which are difficult to calibrate, are examples of nuisance parameters. Before going on to see how the rules of probability can be used to address data analysis problems, let's take a brief look at the history of the subject.

1.4   Some history: Bayes, Laplace and orthodox statistics

About three hundred years ago, people started to give serious thought to the question of how to reason in situations where it is not possible to argue with certainty. James Bernoulli (1713) was perhaps the first to articulate the problem, perceiving the difference between the deductive logic applicable to games of chance and the inductive logic required for everyday life. The open question for him was how the mechanics of the former might help to tackle the inference problems of the latter.

Reverend Thomas Bayes is credited with providing an answer to Bernoulli's question, in a paper published posthumously by a friend (1763). The present-day form of the theorem which bears his name is actually due to Laplace (1812). Not only did Laplace rediscover Bayes' theorem for himself, in far more clarity than did Bayes, he also put it to good use in solving problems in celestial mechanics, medical statistics and even jurisprudence. Despite Laplace's numerous successes, his development of probability theory was rejected by many soon after his death.

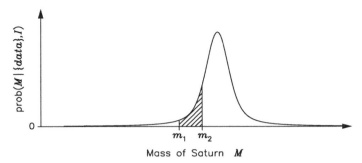

Fig. 1.2 A schematic illustration of the result of Laplace' analysis of the mass of Saturn.

The problem was not really one of substance but concept. To the pioneers such as Bernoulli, Bayes and Laplace, a probability represented a *degree-of-belief* or plausibility: how much they thought that something was true, based on the evidence at hand. To the 19th century scholars, however, this seemed too vague and subjective an idea to be the basis of a rigorous mathematical theory. So they redefined probability as the *long-run relative frequency* with which an event occurred, given (infinitely) many repeated (experimental) trials. Since frequencies can be measured, probability was now seen as an objective tool for dealing with *random* phenomena.

Although the frequency definition appears to be more objective, its range of validity is also far more limited. For example, Laplace used (his) probability theory to estimate the mass of Saturn, given orbital data that were available to him from various astronomical observatories. In essence, he computed the posterior pdf for the mass M, given the data and all the relevant background information I (such as a knowledge of the laws of classical mechanics): $\mathrm{prob}(M\,|\,\{data\},I)$; this is shown schematically in Fig. 1.2. To Laplace, the (shaded) area under the posterior pdf curve between m_1 and m_2 was a measure of how much he believed that the mass of Saturn lay in the range $m_1 \leqslant M < m_2$. As such, the position of the maximum of the posterior pdf represents a best estimate of the mass; its width, or spread, about this optimal value gives an indication of the uncertainty in the estimate. Laplace stated that: ' ... it is a bet of 11,000 to 1 that the error of this result is not 1/100th of its value.' He would have won the bet, as another 150 years' accumulation of data has changed the estimate by only 0.63%! According to the frequency definition, however, we are not permitted to use probability theory to tackle this problem. This is because the mass of Saturn is a constant and not a *random variable*; therefore, it has no frequency distribution and so probability theory cannot be used.

If the pdf of Fig. 1.2 had to be interpreted in terms of the frequency definition, we would have to imagine a large *ensemble* of universes in which everything remains constant apart from the mass of Saturn. As this scenario appears quite far-fetched, we might be inclined to think of Fig. 1.2 in terms of the distribution of the measurements of the mass in many repetitions of the experiment. Although we are at liberty to think about a problem in any way that facilitates its solution, or our understanding of it, having to seek a frequency interpretation for every data analysis problem seems rather perverse.

For example, what do we mean by the 'measurement of the mass' when the data consist of orbital periods? Besides, why should we have to think about many repetitions of an experiment that never happened? What we really want to do is to make the best inference of the mass given the (few) data that we actually have; this is precisely the Bayes and Laplace view of probability.

Faced with the realization that the frequency definition of probability theory did not permit most real-life scientific problems to be addressed, a new subject was invented — *statistics*! To estimate the mass of Saturn, for example, one has to relate the mass to the data through some function called the statistic; since the data are subject to 'random' noise, the statistic becomes the random variable to which the rules of probability theory can be applied. But now the question arises: How should we choose the statistic? The frequentist approach does not yield a natural way of doing this and has, therefore, led to the development of several alternative schools of *orthodox* or *conventional* statistics. The masters, such as Fisher, Neyman and Pearson, provided a variety of different principles, which has merely resulted in a plethora of tests and procedures without any clear underlying rationale. This lack of unifying principles is, perhaps, at the heart of the shortcomings of the cook-book approach to statistics that students are often taught even today.

The frequency definition of probability merely gives the impression of a more objective theory. In reality it just makes life more complicated by hiding the difficulties under the rug, only for them to resurface in a less obvious guise. Indeed, it is not even clear that the concept of 'randomness' central to orthodox statistics is any better-defined than the idea of 'uncertainty' inherent in Bayesian probability theory. For example, we might think that the numbers generated by a call to a function like rand on a computer constitutes a random process: the frequency of the numbers will be distributed uniformly between 0 and 1, and their sequential order will appear haphazard. The illusory nature of this randomness would become obvious, however, if we knew the *algorithm* and the *seed* for the function rand (for then we could predict the sequence of numbers output by the computer). At this juncture, some might argue that, in contrast to our simple illustration above, *chaotic* and *quantum* systems provide examples of physical situations which are intrinsically random. In fact, chaos theory merely underlines the point that we are trying to make: the apparent randomness in the long-term behaviour of a classical system arises because we do not, or cannot, know its initial conditions well enough; the actual temporal evolution is entirely deterministic, and obeys Newton's Second Law of Motion. The quantum case is more difficult to address, since its interpretation (as opposed to its technical success) is still an open question for many people, but we refer the interested reader to Caves *et al.* (2002) for a very insightful viewpoint. Either way, 'randomness' is what we call our inability to predict things which, in turn, reflects our lack of knowledge about the system of interest. This is again consistent with the Bayes and Laplace view of probability, rather than the asserted physical objectivity of the frequentist approach.

To emphasize this last point, that a probability represents a state of knowledge rather than a physically real entity, consider the following example of Jaynes (1989). We are told that a dark bag contains five red balls and seven green ones. If this bag is shaken well, and a ball selected at 'random', then most of us would agree that the probability of drawing a red ball is 5/12 and the probability of drawing a green one is 7/12. If the ball is not returned to the bag, then it seems reasonable that the probability of obtaining a red or green ball on the second draw will depend on the outcome of the first (because there will be one less red or green ball left in the bag). Now suppose that we are not told the outcome of the first draw, but are given the result of the second one. Does the probability of the first draw being red or green change with the knowledge of the second? Initially, many of us would be inclined to say 'no': at the time of the first draw, there were still five red balls and seven green ones in the bag; so, the probabilities for red and green should still be 5/12 and 7/12 irrespective of the outcome of the second draw. The error in this argument becomes obvious if we consider the extreme case of a bag containing only one red and one green ball. Although the second draw cannot affect the first in a physical sense, a knowledge of the second result does influence what we can infer about the outcome of the first one: if the second was green, then the first one must have been red; and vice versa. Thus (conditional) probabilities represent *logical* connections rather than *causal* ones.

The concerns about the subjectivity of the Bayesian view of probability are understandable, and the aim of creating an objective theory is quite laudable. Unfortunately, the frequentist approach does not achieve this goal: neither does its concept of randomness appear very rigorous, or fundamental, under scrutiny and nor does the arbitrariness of the choice of the statistic make it seem objective. In fact, the presumed shortcomings of the Bayesian approach merely reflect a confusion between subjectivity and the difficult technical question of how probabilities should be assigned. The popular argument goes that if a probability represents a degree-of-belief, then it must be subjective because my belief could be different from yours. The Bayesian view is that a probability does indeed represent how much we believe that something is true, but that this belief should be based on all the relevant information available. While this makes the assignment of probabilities an open-ended question, because the information at my disposal may not be the same as that accessible to you, it is not the same as subjectivity. It simply means that probabilities are always conditional, and this conditioning must be stated explicitly. As Jaynes has pointed out, objectivity demands only that two people having the same information should assign the same probability; this principle has played a key rôle in the modern development of the (objective) Bayesian approach.

In 1946, Richard Cox tried to get away from the controversy of the Bayesian versus frequentist view of probability. He decided to look at the question of plausible reasoning afresh, from the perspective of logical consistency. He found that the only rules which met his requirements were those of probability theory. Although the sum and product rules of probability are easy to prove for frequencies (with the aid of a *Venn diagram*), Cox's work shows that their range of validity goes much further. Rather than being restricted to just frequencies, probability theory constitutes the basic calculus for logical and consistent plausible reasoning; for us, that means scientific inference (which is the purpose of data analysis). So, Laplace was right: 'It is remarkable that a science, which commenced with a consideration of games of chance, should be elevated to the rank of the most important subjects of human knowledge.'

1.5   Outline of book

The aim of this book is to show how probability theory can be used directly to address data analysis problems in a straightforward manner. We will start, in Chapter 2, with the simplest type of examples: namely, those involving the estimation of the value of a single parameter. They serve as a good first encounter with Bayes' theorem in action, and allow for a useful discussion about *error-bars* and *confidence intervals*. The examples are extended to two, and then several, parameters in Chapter 3, enabling us to introduce the additional concepts of *correlation* and marginalization. In Chapter 4, we will see how the same principles used for parameter estimation can be applied to the problem of model selection.

Although Cox's work shows that plausibilities, represented by real numbers, should be manipulated according to the rules of probability theory, it does not tell us how to assign them in the first place. We turn to this basic question of assigning probabilities in Chapter 5, where we will meet the important principle of *maximum entropy* (MaxEnt). It may seem peculiar that we leave so fundamental a question to such a late stage, but it is intentional. People often have the impression that Bayesian analysis relies heavily on the use of clever probability assignments and is, therefore, not generally applicable if these are not available. Our aim is to show that even when armed only with a knowledge of pdfs familiar from high school (*binomial*, *Poisson* and *Gaussian*), and naïveté (a *flat*, or *uniform*, pdf), probability theory still provides a powerful tool for obtaining useful results for many data analysis problems. Indeed, we will find that most conventional statistical procedures implicitly assume such elementary assignments. Of course we might do better by thinking deeply about a more appropriate pdf for any given problem, but the point is that it is not usually crucial in practice.

In Chapter 6, we consider *non-parametric* estimation; that is to say, problems where we know so little about the object of interest that we are unable to describe it adequately in terms of a few parameters. Here we will encounter MaxEnt once again, but in the slightly different guise of *image processing*.

In Chapter 7, we focus our attention on the subject of *experimental design*. This concerns the question of 'what are the best data to collect', in contrast to most of this book, which deals with 'what is the optimal way of analysing these data'. This reciprocal question can also be addressed by probability theory, and is of great importance because the benefits of good experimental (or instrumental) design can far outweigh the rewards of the sophisticated analysis of poorer data.

Chapters 8 shows how one of the most widely used data analysis procedures in the physical sciences, *least-squares*, can be extended to deal with more troublesome measurements, such as those with 'outliers'.

Chapters 9 and 10 are concerned with modern numerical techniques for carrying out out Bayesian calculations, when analytical approximations are inadequate and a brute-force approach is impractical. In particular, they provide the first introductory account of the novel idea of *nested sampling*.

Most of the examples used in this book involve continuous pdfs, since this is usually the nature of parameters in real life. As mentioned in Section 1.3, this can simply be considered as the limiting case of an arbitrarily large number of discrete propositions.

Jaynes (2003) correctly warns us, however, that difficulties can sometimes arise if we are not careful in carrying out the limiting procedure explicitly; this is often the underlying cause of so-called paradoxes of probability theory. Such problems also occur in other mathematical calculations, which are quite unrelated to probability. All that one really needs to avoid them is basic professional care allied to common sense. Indeed, we hope that the reader will come to share Laplace's more general view that '*Probability theory is nothing but common sense reduced to calculation*'.

2 Parameter estimation I

Parameter estimation is a common data analysis problem. Like Laplace, for example, we may be interested in knowing the mass of Saturn; or, like Millikan, the charge of the electron. In the simplest case, we are only concerned with the value of a single parameter; such elementary examples are the focus of this chapter. They serve as a good introduction to the use of Bayes' theorem, and allow for a discussion of error-bars and confidence intervals.

2.1 Example 1: is this a fair coin?

Let us begin with the analysis of data from a simple coin-tossing experiment. Suppose I told you that I had been to Las Vegas for my holidays, and had come across a very strange coin in one of the casinos; given that I had observed 4 heads in 11 flips, do you think it was a *fair* coin? By fair, we mean that we would be prepared to lay an even $50:50$ bet on the outcome of a flip being a head or a tail. In ascribing the property of fairness (solely) to the coin we are, of course, assuming that the coin-tosser was not skilled enough to be able to control the initial conditions of the flip (such as the angular and linear velocities). If we decide that the coin was fair, the question which follows naturally is how sure are we that this was so; if it was not fair, how unfair do we think it was?

A sensible way of formulating this problem is to consider a large number of contiguous propositions, or hypotheses, about the range in which the *bias-weighting* of the coin might lie. If we denote the bias-weighting by H, then $H = 0$ and $H = 1$ can represent a coin which produces a tail or a head on every flip, respectively. There is a continuum of possibilities for the value of H between these limits, with $H = 1/2$ indicating a fair coin. The propositions could then be, for example: (a) $0.00 \leqslant H < 0.01$; (b) $0.01 \leqslant H < 0.02$; (c) $0.02 \leqslant H < 0.03$; and so on. Our state of knowledge about the fairness, or the degree of unfairness, of the coin is then completely summarized by specifying how much we believe these various propositions to be true. If we assign a high probability to one (or a closely-grouped few) of these propositions, compared to the others, then this would indicate that we were confident in our estimate of the bias-weighting. If there was no such strong distinction, then it would reflect a high level of ignorance about the nature of the coin.

In the light of the data, and the above discussion, our inference about the fairness of this coin is summarized by the conditional pdf: $\text{prob}(H | \{data\}, I)$. This is, of course, shorthand for the limiting case of a continuum of propositions for the value of H; that is to say, the probability that H lies in an infinitesimally narrow range is given by $\text{prob}(H | \{data\}, I) \, \mathrm{d}H$. To estimate this posterior pdf, we need to use Bayes' theorem (eqn 1.3); it relates the pdf of interest to two others, which are easier to assign:

$$\mathrm{prob}(H\,|\,\{data\},I) \propto \mathrm{prob}(\{data\}\,|\,H,I) \times \mathrm{prob}(H\,|\,I)\,. \tag{2.1}$$

Note that we have omitted the denominator $\mathrm{prob}(\{data\}\,|\,I)$, as it does not involve bias-weighting explicitly, and replaced the equality by a proportionality. If required, we can evaluate the missing constant subsequently from the normalization condition of eqn (1.10):

$$\int_{0}^{1} \mathrm{prob}(H\,|\,\{data\},I)\,\mathrm{d}H = 1\,. \tag{2.2}$$

The prior pdf, $\mathrm{prob}(H\,|\,I)$, on the far right-hand side of eqn (2.1), represents what we know about the coin given only the information I that we are dealing with a 'strange coin from Las Vegas'. Since casinos can be rather dubious places, we should keep a very open mind about the nature of the coin; a simple probability assignment which reflects this is a uniform, or flat, pdf:

$$\mathrm{prob}(H\,|\,I) = \begin{cases} 1 & 0 \leqslant H \leqslant 1, \\ 0 & \text{otherwise}. \end{cases} \tag{2.3}$$

This prior state of knowledge, or ignorance, is modified by the data through the likelihood function: $\mathrm{prob}(\{data\}\,|\,H,I)$. It is a measure of the chance that we would have obtained the data that we actually observed, if the value of the bias-weighting was given (as known). If, in the conditioning information I, we assume that the flips of the coin were independent events, so that the outcome of one did not influence that of another, then the probability of obtaining the data 'R heads in N tosses' is given by the binomial distribution:

$$\mathrm{prob}(\{data\}\,|\,H,I) \propto H^{R}\,(1-H)^{N-R}\,. \tag{2.4}$$

We leave a formal derivation of this pdf to Chapter 5, but point out that eqn (2.4) seems reasonable because H is the chance of obtaining a head on any flip, and there were R of them, and $1-H$ is the corresponding probability for a tail, of which there were $N-R$. For simplicity, an equality has again been replaced by a proportionality; this is permissible since the omitted terms contain factors of only R and N, rather than the quantity of interest H.

According to eqn (2.1), the product of eqns (2.3) and (2.4) yields the posterior pdf we require; it represents our state of knowledge about the nature of the coin in the light of the data. To get a feel for this result, it is instructive to see how this pdf evolves as we obtain more and more data pertaining to the coin. This is done with the aid of data generated in a computer simulation, and the results of their analyses is shown in Fig. 2.1. The panel in the top left-hand corner shows the posterior pdf for H given no data; it is, of course, the same as the prior pdf of eqn (2.3). It indicates that we have no more reason to believe that the coin is fair than we have to think that it is double-headed, double-tailed, or of any other intermediate bias-weighting.

Suppose that the coin is flipped once and it comes up heads; what can we now say about the value of H? The resulting posterior pdf, shown in the second panel of Fig. 2.1,

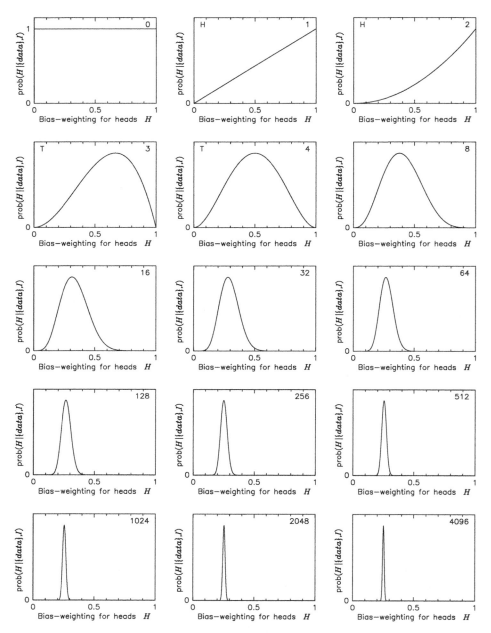

Fig. 2.1 The evolution of the posterior pdf for the bias-weighting of a coin, $\mathrm{prob}(H\,|\,\{data\}, I)$, as the number of data available increases. The figure on the top right-hand corner of each panel shows the number of data analysed; in the early panels, the H or T in the top left-hand corner shows whether the result of that (last) flip was a head or a tail.

goes to zero for $H = 0$ and rises linearly to having the greatest value at $H = 1$. Based purely on this single datum, it is most probable that the (strange) coin has two heads; after all, we don't have any empirical evidence that it even has a tail yet. Although $H = 1$ is in some ways our 'best' estimate so far, the posterior pdf indicates that this value is not much more probable than many others. The only thing we can really be sure about is that the coin is not double-tailed; hence, the posterior pdf is zero at $H = 0$.

If the coin is flipped for a second time and again comes up heads, the posterior pdf becomes slightly more peaked towards $H = 1$ (proportional to H^2); this is plotted in the third panel of Fig. 2.1. Since we still haven't seen a tail, our inclinations following the first datum are just reinforced. As soon as a tail is obtained, however, the posterior pdf for $H = 1$ also drops to zero; we are then sure that the coin is not double-headed either. The next panel shows the resultant pdf, given that the third flip produced a tail (proportional to $(1 - H) H^2$). If the fourth flip also comes up tails, then the maximum of the posterior pdf is at $H = 0.5$. It then becomes most probable that the coin is fair, but there is still a large degree of uncertainty in this estimate; this is indicated graphically in panel 5 of Fig. 2.1.

The remainder of Fig. 2.1 shows how the posterior pdf for H evolves as the number of data analysed becomes larger and larger. We see that the position of the maximum wobbles around, but that the amount by which it does so decreases with the increasing number of observations. The width of the posterior pdf also becomes narrower with more data, indicating that we are becoming increasingly confident in our estimate of the bias-weighting. For the coin in this example, the best estimate of H eventually converges to 0.25. This was, of course, the value chosen to simulate the flips; they can be thought of as coming from a tetrahedral coin with a head on one face and tails on the other three!

2.1.1 Different priors

Most people tend to be happy with the binomial distribution for the likelihood function, but worry about the prior pdf. The uniform assignment of eqn (2.3) was chosen mostly for its simplicity; it is just a naïve way of encoding a lot of initial ignorance about the coin. How would our inference about the fairness of the coin have changed if we had chosen a different prior?

To address this question, let's repeat the analysis of the data above with two alternative prior pdfs; the results are shown in Fig. 2.2. The solid line is for the uniform pdf used in Fig. 2.1, and is included for ease of comparison. One of the alternative priors is peaked around $H = 0.5$ and reflects our background information that most coins are fair (even in Las Vegas); it is plotted with a dashed line. It has a width which is broad enough to comfortably accommodate the possibility that H might be as low as 0.35, or as high as 0.65, but expresses considerable doubt about greater levels of biased behaviour. The second alternative, shown with a dotted line, is very sharply peaked at $H = 0$ and $H = 1$; it indicates that we would expect a 'strange coin from a casino' to be heavily biased, one way or another.

Figure 2.2 shows how the posterior pdfs for the different priors evolve, as more and more data become available; they have all been scaled vertically to have the same

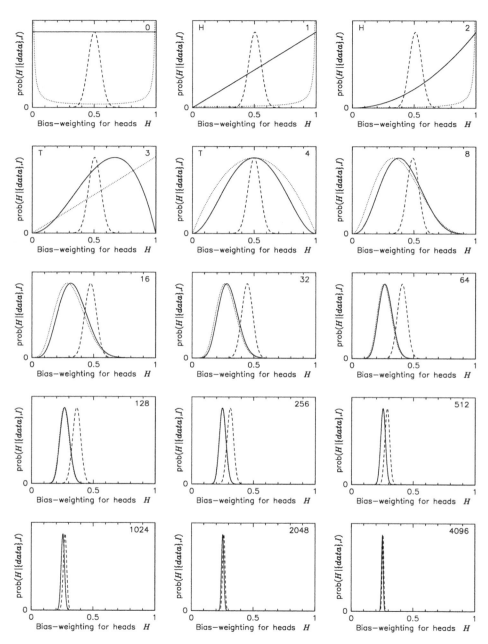

Fig. 2.2 The effect of different priors, prob($H|I$), on the posterior pdf for the bias-weighting of a coin. The solid line is the same as in Fig. 2.1, and is included for ease of comparison. The case for two alternative priors, reflecting slightly different assumptions in the conditioning information I, are shown with dashed and dotted lines.

maximum value to aid comparison between them. We find that when there are few data, the resulting posteriors are different in detail; as the number of data increases, they all become more sharply peaked and converge to the same answer. This seems quite reasonable. The outcome of only a few flips tells us little about the fairness of the coin. Our state of knowledge after the analysis of these data is, therefore, strongly dependent on what we knew or assumed before the results; hence, the posteriors are somewhat different. As the empirical evidence grows, we are eventually led to the same conclusions irrespective of our initial beliefs; the posterior pdf is then dominated by the likelihood function, and the choice of the prior becomes largely irrelevant.

Two further curious points may be noted: (i) it takes quite a lot of flips to be able to estimate the bias-weighting with some degree of confidence (about a thousand to pin it down between 0.2 and 0.3); (ii) the posterior pdfs for the solid and dotted lines converge together quite quickly, but the dashed case takes much longer. The answer to the first observation is that it just does, but the number of flips required will depend on the actual bias-weighting of the coin. If the coin was tossed ten times and came up tails on every occasion, we would rapidly conclude that it was biased; on the other hand, the result of 45 heads and 55 tails in a hundred flips would still leave us somewhat uncertain as to whether the coin was fair. With regard to the second point, both the flat (solid) and the spiky (dotted) priors encode a large degree of ignorance about the nature of the coin. Despite the peculiar shape of the latter, it is fairly flat for most values of H; the strange-looking spikes at $H = 0$ and $H = 1$ disappear as soon as a head and a tail have been observed. The 'fair-minded' prior (dashed line), however, claims to be moderately well-informed about the character of the coin regardless of the data. It, therefore, takes much more to be convinced that the coin is not fair. Even though the prior probability for $H = 0.5$ was about a million times greater than that of $H = 0.25$, a thousand flips were enough to drag it (kicking and screaming, perhaps) to the conclusion that the value of H was less than 0.3 (but more than 0.2).

2.1.2 Sequential or one-step data analysis?

If we have a set of data $\{D_k\}$ comprising the outcome of N flips of a coin ($k = 1, 2, \ldots, N$), then Bayes' theorem tells us that the posterior pdf for H is given by

$$\mathrm{prob}(H\,|\,\{D_k\}, I) \propto \mathrm{prob}(\{D_k\}\,|\,H, I) \times \mathrm{prob}(H\,|\,I)\,. \qquad (2.5)$$

This is a one-step process in that we consider the data collectively, as a whole. As an alternative, we could also think of analysing the data sequentially (as they arrive). That is to say, we start by computing the posterior pdf based on the first datum D_1, $\mathrm{prob}(H\,|\,D_1, I)$, and use it as the prior for the analysis of the second datum D_2; the new posterior could then be used as the prior for the third datum, and so on. If we continue this procedure, will we obtain the same result as the one-step approach?

To simplify matters, let us consider the case of just two data. The posterior pdf for H, based on both, is merely a special case of eqn (2.5) for $N = 2$:

$$\mathrm{prob}(H\,|\,D_2, D_1, I) \propto \mathrm{prob}(D_2, D_1\,|\,H, I) \times \mathrm{prob}(H\,|\,I)\,. \qquad (2.6)$$

Equally well, we could use Bayes' theorem to express the posterior in terms of pdfs conditional on D_1 throughout (like the background information I):

$$\mathrm{prob}(H\,|\,D_2, D_1, I) \,\propto\, \mathrm{prob}(D_2\,|\,H, D_1, I) \times \mathrm{prob}(H\,|\,D_1, I)\,. \qquad (2.7)$$

This shows that the prior in eqn (2.6) can certainly be replaced by the posterior pdf based on the first datum, but the other term doesn't quite look like the likelihood function for the second datum. Implicit in the assignment of the binomial pdf of eqn (2.4), however, was the assumption (subsumed in I) that the data were *independent*. This means that, given the value of H, the result of one flip does not influence what we can infer about the outcome of another; mathematically, we write this as

$$\mathrm{prob}(D_2\,|\,H, D_1, I) \,=\, \mathrm{prob}(D_2\,|\,H, I)\,.$$

Substituting this in eqn (2.7) yields the desired relationship, showing that the two data can either be analysed together or one after the other. This argument can be extended to the third datum D_3, giving

$$\mathrm{prob}(H\,|\,D_3, D_2, D_1, I) \,\propto\, \mathrm{prob}(D_3\,|\,H, I) \times \mathrm{prob}(H\,|\,D_2, D_1, I)\,,$$

and repeated until all the data have been included. We should not really be surprised that both the one-step and sequential methods of analysis give the same answer; after all, the requirements of such consistency was what led Cox to the rules of probability theory in the first place.

When one first learns about Bayes' theorem, and sees how the data modify the prior through the likelihood function, there is occasionally a temptation to use the resulting posterior pdf as the prior for a re-analysis of the same data. It would be erroneous to do this, and the results quite misleading. In order to justify any data analysis procedure, we must be able to relate the pdf of interest to others used in its calculation through the sum and product rules of probability (or their corollaries). If we cannot do this then the analysis will be suspect at best, and open to logical inconsistencies. For the case of this proposed 'boot-strapping' we would, in fact, be trying to relate the posterior pdf to itself; this can only be done by an equality, and not through the likelihood function. If we persist in our folly, and keep repeating it, the resulting posterior pdfs will become sharper and sharper; we will just fool ourselves into thinking that the quantity of interest can be estimated far more accurately than is warranted by the data.

2.2   Reliabilities: best estimates, error-bars and confidence intervals

We have seen how the posterior pdf encodes our inference about the value of a parameter, given the data and the relevant background information. Often, however, we wish to summarize this with just two numbers: the best estimate and a measure of its reliability. Since the probability (density) associated with any particular value of the parameter is a measure of how much we believe that it lies in the neighbourhood of that point, our best estimate is given by the maximum of the posterior pdf. If we denote the quantity of interest by X, with a posterior pdf $P = \mathrm{prob}(X\,|\,\{data\}, I)$, then the best estimate of its value X_0 is given by the condition

$$\left. \frac{\mathrm{d} P}{\mathrm{d} X} \right|_{X_\mathrm{O}} = 0 . \tag{2.8}$$

Strictly speaking, we should also check the sign of the *second derivative* to ensure that X_O represents a maximum rather than a minimum (or a point of inflexion):

$$\left. \frac{\mathrm{d}^2 P}{\mathrm{d} X^2} \right|_{X_\mathrm{O}} < 0 . \tag{2.9}$$

In writing the derivatives of P with respect to X we are, of course, assuming that X is a continuous parameter. If it could only take discrete values, our best estimate would still be that which gave the greatest posterior probability; it's just that we couldn't then use the calculus notation of eqns (2.8) and (2.9), because the *gradients* are not defined (since the increment in X cannot be infinitesimally small).

To obtain a measure of the reliability of this best estimate, we need to look at the width or spread of the posterior pdf about X_O. When considering the behaviour of any function in the neighbourhood of a particular point, it is often helpful to carry out a *Taylor series* expansion; this is simply a standard tool for (locally) approximating a complicated function by a low-order *polynomial*. Rather than dealing directly with the posterior pdf P, which is a 'peaky' and positive function, it is better to work with its *logarithm L*,

$$L = \log_\mathrm{e} \left[\, \mathrm{prob}(X \,|\, \{data\}, I) \, \right] , \tag{2.10}$$

since this varies much more slowly with X. Expanding L about the point $X = X_\mathrm{O}$, we have

$$L = L(X_\mathrm{O}) + \frac{1}{2} \left. \frac{\mathrm{d}^2 L}{\mathrm{d} X^2} \right|_{X_\mathrm{O}} (X - X_\mathrm{O})^2 + \cdots , \tag{2.11}$$

where the best estimate of X is given by the condition

$$\left. \frac{\mathrm{d} L}{\mathrm{d} X} \right|_{X_\mathrm{O}} = 0 , \tag{2.12}$$

which is equivalent to eqn (2.8) because L is a *monotonic* function of P.

The first term in the Taylor series, $L(X_\mathrm{O})$, is a constant and tells us nothing about the shape of the posterior pdf. The linear term, which would be proportional to $X - X_\mathrm{O}$, is missing because we are expanding about the maximum (as indicated by eqn 2.12). The *quadratic* term is, therefore, the dominant factor determining the width of the posterior pdf and plays a central rôle in the reliability analysis. Ignoring all the higher-order contributions, the exponential of eqn (2.11) yields

$$\mathrm{prob}(X \,|\, \{data\}, I) \approx A \, \exp \left[\frac{1}{2} \left. \frac{\mathrm{d}^2 L}{\mathrm{d} X^2} \right|_{X_\mathrm{O}} (X - X_\mathrm{O})^2 \right] , \tag{2.13}$$

where A is a normalization constant. Although this expression looks a little weird, what we have really done is to approximate the posterior pdf by the ubiquitous Gaussian distribution. Also known as the *normal* distribution, it is usually written as

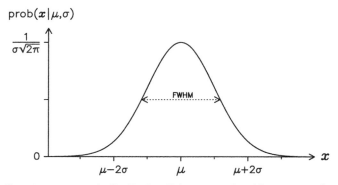

Fig. 2.3 The Gaussian, or normal, distribution. It is symmetric with respect to the maximum, at $x = \mu$, and has a full width at half maximum (FWHM) of about $2.35\,\sigma$.

$$\mathrm{prob}(x|\mu,\sigma) \;=\; \frac{1}{\sigma\sqrt{2\pi}}\;\exp\!\left[-\frac{(x-\mu)^2}{2\,\sigma^2}\right], \qquad (2.14)$$

and is plotted in Fig. 2.3; this function is symmetric about the maximum, at $x = \mu$, and has a width which is proportional to σ. Comparing the exponents of eqns (2.13) and (2.14), we are reassured to find that the posterior pdf for X has a maximum at $X = X_0$; its width, characterized by the parameter σ, is inversely related to the square root of the second derivative of L at $X = X_0$:

$$\sigma \;=\; \left(-\left.\frac{\mathrm{d}^2 L}{\mathrm{d}X^2}\right|_{X_0}\right)^{-1/2}, \qquad (2.15)$$

where $\mathrm{d}^2 L/\mathrm{d}X^2$ is necessarily negative following eqn (2.9). Our inference about the quantity of interest is conveyed very concisely, therefore, by the statement

$$X = X_0 \pm \sigma, \qquad (2.16)$$

where X_0 is the best estimate of the value of X, and σ is a measure of its reliability; the parameter σ is usually referred to as the *error-bar*.

The expression of eqn (2.16) is, of course, just shorthand for the Gaussian approximation of eqn (2.13). The integral properties of the normal distribution tell us that the probability that the true value of X lies within $\pm 1\sigma$ of $X = X_0$ is 67%:

$$\mathrm{prob}(X_0 - \sigma \leqslant X < X_0 + \sigma\,|\,\{data\}, I) = \int\limits_{X_0-\sigma}^{X_0+\sigma} \mathrm{prob}(X, \{data\}|I)\,\mathrm{d}X \approx 0.67.$$

Similarly, the probability that X lies within $\pm 2\sigma$ of X_0 is 95%; we would be quite surprised, however, if our best estimate of X was wrong by more than about 3σ.

2.2.1   The coin example

As a concrete example of the above analysis, let's consider the case of the coin-tossing experiment of the previous section. Multiplying the uniform prior of eqn (2.3) with the binomial likelihood function of eqn (2.4), we obtain the posterior pdf for the bias-weighting:

$$\text{prob}(H|\{data\}, I) \propto H^R (1-H)^{N-R},$$

where $0 \leqslant H \leqslant 1$. Taking its natural logarithm, according to eqn (2.10), gives

$$L = \text{constant} + R \log_e(H) + (N-R) \log_e(1-H). \tag{2.17}$$

For the best estimate of H and its error-bar, we need both the first and the second derivative of L; differentiating twice, with respect to H, we have

$$\frac{dL}{dH} = \frac{R}{H} - \frac{(N-R)}{(1-H)} \quad \text{and} \quad \frac{d^2L}{dH^2} = -\frac{R}{H^2} - \frac{(N-R)}{(1-H)^2}. \tag{2.18}$$

However, eqn (2.12) states that, at the optimal value for the bias-weighting H_O, the first derivative is equal to zero:

$$\frac{dL}{dH}\bigg|_{H_O} = \frac{R}{H_O} - \frac{(N-R)}{(1-H_O)} = 0.$$

Hence, algebraic rearrangement of this equation tells us that the best estimate of the bias-weighting is given by the relative frequency of outcomes of heads:

$$H_O = \frac{R}{N}. \tag{2.19}$$

According to eqn (2.15), the associated error-bar is related to the second derivative of L evaluated at $H = H_O$. Substituting for R from eqn (2.19), into the right-hand formula in eqn (2.18), and after some rearrangement and cancellation, we obtain

$$\frac{d^2L}{dH^2}\bigg|_{H_O} = -\frac{N}{H_O(1-H_O)}.$$

Equation (2.15) then tells us that the error-bar for the estimate of the bias-weighting is given by

$$\sigma = \sqrt{\frac{H_O(1-H_O)}{N}}. \tag{2.20}$$

Since H_O does not vary a lot after a moderate amount of data have been analysed, the numerator tends to a constant value; thus the width of the posterior becomes inversely proportional to the square root of the number of data, as can be seen in Fig. 2.1. This formula for the error-bar also confirms our earlier assertion that it is easier to identify a highly biased coin than it is to be confident that it's fair (because the numerator in eqn (2.20) has its greatest value when $H_O = 0.5$). The posterior pdf for H, given the data 9 heads in 32 flips, and the Gaussian approximation to it resulting from the above analysis ($H = 0.28 \pm 0.08$), are shown in Fig. 2.4.

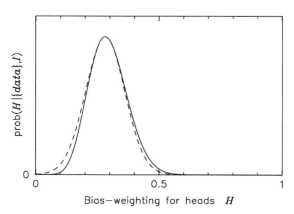

Fig. 2.4 The Gaussian, or quadratic, approximation (dashed line) to the posterior pdf for the bias-weighting of the coin, given 9 heads in 32 flips.

2.2.2 Asymmetric posterior pdfs

The preceding analysis, leading to the error-bar formula of eqn (2.15), relies on the validity of the quadratic expansion of eqn (2.11). Figure 2.4 illustrates that this is usually a reasonable approximation; in fact, the agreement with the Gaussian pdf tends to improve steadily as the number of data increases. There are times, however, when the posterior pdf is markedly asymmetric; this can be seen in the early panels of Figs. 2.1 and 2.2. While the maximum of the posterior (X_O) can still be regarded as giving the best estimate, the true value is now more likely to be on one side of this rather than the other. The concept of an error-bar does not seem appropriate in this case, as it implicitly entails the idea of symmetry.

A good way of expressing the reliability with which a parameter can be inferred, for an asymmetric posterior pdf, is through a *confidence interval*. Since the area under the posterior pdf between X_1 and X_2 is proportional to how much we believe that X lies in that range, the *shortest interval* that encloses 95% of the area represents a sensible measure of the uncertainty of the estimate. Assuming that the posterior pdf has been normalized, to have unit area, we need to find X_1 and X_2 such that

$$\text{prob}(X_1 \leqslant X < X_2 \,|\, \{data\}, I) = \int_{X_1}^{X_2} \text{prob}(X \,|\, \{data\}, I) \, dX \approx 0.95 \,, \quad (2.21)$$

where the difference $X_2 - X_1$ is as small as possible. The region $X_1 \leqslant X < X_2$ is called the shortest 95% confidence interval, and is illustrated in Fig. 2.5.

A natural question which arises is 'Why have we chosen the 95% confidence level?' The answer is that this is traditionally seen as a reasonable value, as it provides a respectably conservative estimate of the reliability. There is nothing to stop us from giving the shortest 50%, 70%, 99%, or any other, confidence interval. Indeed, there is something to be said for listing a whole set of nested intervals since this provides a more

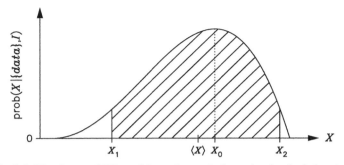

Fig. 2.5 The shortest 95% confidence interval, shown by the shaded region.

complete picture of the reliability analysis; by doing so, however, we are merely recon-structing the posterior pdf!

When dealing with a highly asymmetric posterior pdf, the question of what we mean by the 'best' estimate is rather open-ended. The maximum still indicates the single most probable value for the parameter of interest, but the *mean*, or *expectation*, can be thought of as being more representative as it takes into account the skewness of the pdf. For a normalized posterior, this weighted average $\langle X \rangle$ is given by

$$\langle X \rangle = \int X \, \text{prob}(X|\{data\}, I) \, \text{d}X \,, \qquad (2.22)$$

and is sometimes denoted by $E(X)$ or $\overline{X}$ (the latter not to be confused with 'not X'); as usual, the integral is replaced by a sum when X can only take discrete values. If the posterior pdf has not been normalized then, of course, the right-hand side must be divided by $\int \text{prob}(X|\{data\}, I) \, \text{d}X$. For the case of the pdf shown in Fig. 2.5, the mean lies slightly to the left of X_0. If the posterior pdf is symmetric about the maximum, as in the Gaussian distribution, then the mean and the maximum become coincident ($\langle X \rangle = X_0$).

2.2.3 Multimodal posterior pdfs

So far, we have only considered posterior pdfs which have a single maximum. Depend-ing on the nature of the experimental data to be analysed, we can sometimes obtain posteriors which are *multimodal*; this is depicted schematically in Fig. 2.6. There is no difficulty when one of the maxima is very much larger than the others: we can sim-ply ignore the subsidiary solutions, to a good approximation, and concentrate on the global maximum. The problem arises when there are several maxima of comparable magnitude. What do we now mean by a best estimate, and how should we quantify its reliability?

Well, to a large extent, the difficulty is of our own making. The posterior pdf gives a complete description of what we can infer about the value of the parameter in the light of the data, and our relevant prior knowledge. The idea of a best estimate and an error-bar, or even a confidence interval, is merely an attempt to summarize the posterior with just two or three numbers; sometimes this just can't be done, and so these concepts

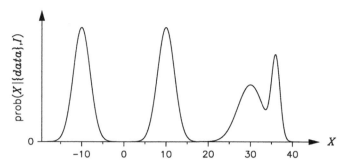

Fig. 2.6 A multimodal posterior pdf. Since its shape cannot be summarized by just a couple of numbers, the concept of a best estimate and an error-bar is inappropriate.

are not valid. The posterior pdf still exists, and we are free to draw from it whatever conclusions are appropriate.

Consider the special case of a multimodal distribution having two (roughly) equal-sized probability bumps; such a *bimodal* pdf is seen in Fig. 2.6, if the structure to the right of $X = 20$ is ignored. This posterior conveys the message that the value of X is either -10, give or take a little bit, or about $+10$; we could write it as: '$X = -10 \pm 2$ or $X = +10 \pm 2$'. Since the mean of the posterior pdf is still unique, it is sometimes suggested that this should be used as a (single) best estimate. The problem with that argument is that the expectation $\langle X \rangle \approx 0$, a value which the posterior pdf indicates is very improbable! Even if we turned a blind eye to this difficulty, and used the mean as the best estimate, we would have to assign a sizeable error-bar to encompass the most probable values of X; this would again be somewhat misleading, as it is not a good reflection of the information inherent in the posterior pdf. For the bimodal case we can characterize the posterior in terms of a few numbers: two best estimates and their associated error-bars, or disjoint confidence intervals. For a general multimodal pdf, the most honest thing we can do is just display the posterior itself.

2.3 Example 2: Gaussian noise and averages

For our second example, let us consider the problem of estimating the mean of a Gaussian process. The normal distribution, which was defined in eqn (2.14), and plotted in Fig. 2.3, is often used as a theoretical model for describing the *noise* (or imperfections) associated with experimental data. We leave its formal derivation with maximum entropy to Chapter 5, but note that its use is traditionally justified by appealing to the *central limit* theorem. For our present purposes, we need merely say that the probability of the kth datum having a value x_k is given by

$$\text{prob}(x_k|\mu,\sigma) = \frac{1}{\sigma\sqrt{2\pi}} \exp\left[-\frac{(x_k-\mu)^2}{2\sigma^2}\right],\qquad (2.23)$$

where μ is the true value of the parameter of interest, and σ is a measure of the error in its measurement. Given a set of data $\{x_k\}$, what is the best estimate of μ and how confident are we of this prediction?

In this example, we will restrict ourselves to the relatively easy case when the value of σ is known; the more realistic situation, when this condition is relaxed, is left to Section 3.3. Our inference about the value of μ is expressed, therefore, by the posterior pdf $\text{prob}(\mu|\{x_k\}, \sigma, I)$; to help us calculate it, we use Bayes' theorem:

$$\text{prob}(\mu|\{x_k\}, \sigma, I) \propto \text{prob}(\{x_k\}|\mu, \sigma, I) \times \text{prob}(\mu|\sigma, I). \qquad (2.24)$$

Although a knowledge of σ could be subsumed into the background information I, we will continue to state this conditioning explicitly (for future reference). If we assume that the data are independent, so that the measurement of one datum does not influence what we can infer about the outcome of another (when given the values of μ and σ), then the likelihood function is given by the product of the probabilities of obtaining the N individual data:

$$\text{prob}(\{x_k\}|\mu, \sigma, I) = \prod_{k=1}^{N} \text{prob}(x_k|\mu, \sigma, I). \qquad (2.25)$$

This follows from the repeated use of eqn (1.2), which reduces to

$$\text{prob}(X, Y|I) = \text{prob}(X|I) \times \text{prob}(Y|I),$$

since the notion of independence implies that

$$\text{prob}(X|Y, I) = \text{prob}(X|I).$$

As a knowledge of the width of a Gaussian tells us nothing about the position of its centre, let us assign a simple uniform pdf for the prior:

$$\text{prob}(\mu|\sigma, I) = \text{prob}(\mu|I) = \begin{cases} A & \mu_{\min} \leqslant \mu \leqslant \mu_{\max}, \\ 0 & \text{otherwise}, \end{cases} \qquad (2.26)$$

where the normalization constant A is given by the reciprocal of the range $\mu_{\max} - \mu_{\min}$, determined by the relevant background information I. According to eqn (2.24), if we multiply this prior by the likelihood function resulting from eqns (2.23) and (2.25), we obtain the following for the logarithm of the posterior pdf L:

$$L = \log_e\left[\text{prob}(\mu|\{x_k\}, \sigma, I)\right] = \text{constant} - \sum_{k=1}^{N} \frac{(x_k - \mu)^2}{2\sigma^2}, \qquad (2.27)$$

where the constant includes all terms not involving μ; the posterior pdf is, of course, zero outside the range $\mu_{\min}$ to $\mu_{\max}$.

To find the best estimate μ_o, eqn (2.12) tells us that we need to differentiate L with respect to μ and set the derivative to zero:

$$\left.\frac{\mathrm{d}L}{\mathrm{d}\mu}\right|_{\mu_\mathrm{o}} = \sum_{k=1}^{N}\frac{x_k-\mu_\mathrm{o}}{\sigma^2} = 0\,.$$

Since σ is a constant which is independent of k, we can take it outside the summation sign and rearrange the equation as follows:

$$\sum_{k=1}^{N}x_k = \sum_{k=1}^{N}\mu_\mathrm{o} = N\mu_\mathrm{o}\,.$$

Thus the best estimate of the value of μ is given by the simple arithmetic average of the sample of measurements $\{x_k\}$:

$$\mu_\mathrm{o} = \frac{1}{N}\sum_{k=1}^{N}x_k\,, \qquad (2.28)$$

irrespective of the magnitude of the error in the measurement process. Our confidence in the reliability of this best estimate, however, does depend on the size of σ, as indicated by the second derivative of L:

$$\left.\frac{\mathrm{d}^2L}{\mathrm{d}\mu^2}\right|_{\mu_\mathrm{o}} = -\sum_{k=1}^{N}\frac{1}{\sigma^2} = -\frac{N}{\sigma^2}\,.$$

According to eqn (2.15), the error-bar we should associate with μ_o is given by the square root of the inverse of (minus) this quantity; thus we can summarize our inference about the value of μ by the statement

$$\mu = \mu_\mathrm{o} \pm \frac{\sigma}{\sqrt{N}}\,. \qquad (2.29)$$

Just as in the case of the coin-tossing experiment, we encounter the familiar result that the reliability of the estimate is proportional to the square root of the number of data.

   In Section 2.2, we noted that the error-bar analysis relies on the validity of the quadratic expansion of eqn (2.11). For the case of Gaussian noise considered above, this is not only a reasonable approximation but an exact identity (because all higher derivatives of L are zero). Thus, the posterior pdf is completely defined by the parameters of eqn (2.29). The only proviso concerns the values of μ_min and μ_max. We could, in principle, express an inordinate amount of prior ignorance by letting them tend to $\pm\infty$ (although not at the outset!) but, as long as the range they span is large enough, their actual value has no effect on the posterior pdf. If the best estimate and error-bar allow for values of μ outside the range μ_min to μ_max, then it is most honest simply to display the posterior itself with its cut-offs; in that case, probability theory would just be warning us that our prior knowledge was at least as important as the data.

2.3.1 Data with different-sized error-bars

In the above analysis, we assumed that the magnitude of the error-bar for each datum was the same; this may well be reasonable if all the measurements were made with a given experimental set-up. Suppose, however, that the data were obtained from several laboratories using equipment of varying sophistication. How should we then combine the evidence from observations of differing quality?

Let us assume that the measurement error can still be modelled through a Gaussian pdf, so that the probability of the kth datum having a value x_k is

$$\text{prob}(x_k|\mu,\sigma_k) = \frac{1}{\sigma_k\sqrt{2\pi}} \exp\left[-\frac{(x_k-\mu)^2}{2\,\sigma_k^2}\right],$$

where the error-bar for each datum σ_k need not be of the same size. Following the arguments between eqns (2.23) and (2.27), we find that the logarithm of the posterior pdf for μ is now given by

$$L = \log_e\left[\text{prob}(\mu|\{x_k\},\{\sigma_k\},I)\right] = \text{constant} - \sum_{k=1}^{N}\frac{(x_k-\mu)^2}{2\,\sigma_k^2}.$$

As usual, we obtain the expression for the best estimate by setting the first derivative of L equal to zero; since the resulting equation does not simplify as much as before, the formula for μ_o is somewhat more complicated:

$$\mu_o = \sum_{k=1}^{N} w_k x_k \bigg/ \sum_{k=1}^{N} w_k, \qquad \text{where } w_k = \frac{1}{\sigma_k^2}. \tag{2.30}$$

Rather than the best estimate being given by the arithmetic mean of the data, we must now calculate their weighted average. Less reliable data will have larger error-bars σ_k and correspondingly smaller weights w_k. The second derivative of L yields the error-bar for the best estimate and allows us to summarize our inference about μ as

$$\mu = \mu_o \pm \left(\sum_{k=1}^{N} w_k\right)^{-1/2}. \tag{2.31}$$

Note that if all the data were of comparable quality, so that $\sigma_k=\sigma$, then eqns (2.30) and (2.31) would reduce to the simpler forms of eqns (2.28) and (2.29).

2.4 Example 3: the lighthouse problem

For the third example, we follow Gull (1988) in considering a very instructive problem found on a problems sheet for first-year undergraduates at Cambridge: 'A lighthouse is somewhere off a piece of straight coastline at a position α along the shore and a distance β out at sea. It emits a series of short highly collimated flashes at random intervals and hence at random azimuths. These pulses are intercepted on the coast by photo-detectors

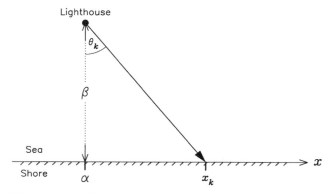

Lighthouse

θ_k

β

Sea

Shore

α

x_k

x

Fig. 2.7 A schematic illustration of the geometry of the lighthouse problem.

that record only the fact that a flash has occurred, but not the angle from which it came. N flashes have so far been recorded at positions $\{x_k\}$. Where is the lighthouse?'

The geometry of the problem above is illustrated in Fig. 2.7. Given the nature of the lighthouse emissions, it seems reasonable to assign a uniform pdf for the azimuth of the kth datum θ_k:

$$\mathrm{prob}(\theta_k | \alpha, \beta, I) = \frac{1}{\pi}, \qquad (2.32)$$

where the angle must lie between $\pm \pi/2$ radians (or $\pm 90°$) to have been detected. Since the photo-detectors are only sensitive to position along the coast rather than direction, we must relate θ_k to x_k. An inspection of Fig. 2.7, and elementary *trigonometry*, allows us to write:

$$\beta \tan \theta_k = x_k - \alpha. \qquad (2.33)$$

As we shall see in Section 3.6, when dealing with the subject of *changing variables*, the relationship of eqn (2.33) can be used to rewrite the pdf of eqn (2.32) as

$$\mathrm{prob}(x_k | \alpha, \beta, I) = \frac{\beta}{\pi \left[\beta^2 + (x_k - \alpha)^2 \right]}. \qquad (2.34)$$

This tells us that the probability that the kth flash will be recorded at position x_k, knowing the coordinates of the lighthouse (α, β), is given by a *Cauchy* distribution. The functional form of this pdf is met frequently in physics and is often called a *Lorentzian*. It is symmetric about the maximum, at $x_k = \alpha$, and has a FWHM of 2β; the distribution is plotted in Fig. 2.8.

Inferring the position of the lighthouse from the data involves the estimation of both α and β; as a two-parameter problem, this is outside the scope of the present chapter. We will therefore assume that the distance out to sea is known and reduce it to a single parameter example; the reader should have no difficulty, however, in tackling the full problem after the material of the next chapter. Our inference about the position

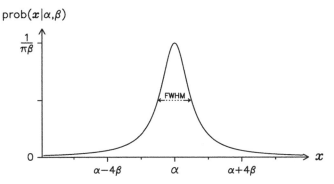

prob($x|\alpha,\beta$)

Fig. 2.8 The Cauchy, or Lorentzian, distribution. It is symmetric with respect to the maximum, at $x = \alpha$, and has a FWHM of twice β.

of the lighthouse is then expressed by the posterior pdf $\mathrm{prob}(\alpha|\{x_k\},\beta,I)$; to help us calculate it, we begin by writing down Bayes' theorem:

$$\mathrm{prob}(\alpha|\{x_k\},\beta,I) \propto \mathrm{prob}(\{x_k\}|\alpha,\beta,I) \times \mathrm{prob}(\alpha|\beta,I)\,. \qquad (2.35)$$

As a knowledge of β, without the data, tells us nothing new about the position of the lighthouse along the shore, let us assign a simple uniform pdf for the prior:

$$\mathrm{prob}(\alpha|\beta,I) = \mathrm{prob}(\alpha|I) = \begin{cases} A & \alpha_{\min} \leqslant \alpha \leqslant \alpha_{\max}\,, \\ 0 & \text{otherwise}\,, \end{cases} \qquad (2.36)$$

where $\alpha_{\min}$ and $\alpha_{\max}$ could represent the limits of the coastline, if these were known, or be made arbitrarily large (up to the size of the Earth!) if we were very ignorant; the normalization constant A is, of course, just the reciprocal of this prior range. Since the recording of a signal at one photo-detector does not influence what we can infer about the position of another measurement (when given the location of the lighthouse), the likelihood function for these independent data is just the product of the probabilities for obtaining the N individual detections:

$$\mathrm{prob}(\{x_k\}|\alpha,\beta,I) = \prod_{k=1}^{N} \mathrm{prob}(x_k|\alpha,\beta,I)\,. \qquad (2.37)$$

Substituting the prior of eqn (2.36) and the likelihood function resulting from eqns (2.34) and (2.37) into Bayes' theorem of eqn (2.34), allows us to write the logarithm of the posterior pdf as

$$L = \log_{e}\big[\mathrm{prob}(\alpha|\{x_k\},\beta,I)\big] = \text{constant} - \sum_{k=1}^{N} \log_{e}\big[\beta^2 + (x_k - \alpha)^2\big]\,, \qquad (2.38)$$

where the constant includes all terms not involving α. For the purposes of this example, we will assume that the prior range is so large that we need not worry about any cut-offs imposed on the posterior.

The best estimate of the position α_o is given by the maximum of the posterior pdf; differentiating eqn (2.38) once, with respect to α, we obtain the condition

$$\left.\frac{\mathrm{d}L}{\mathrm{d}\alpha}\right|_{\alpha_o} = 2 \sum_{k=1}^{N} \frac{x_k - \alpha_o}{\beta^2 + (x_k - \alpha_o)^2} = 0 \,. \qquad (2.39)$$

Unfortunately, it is difficult to rearrange this equation so that α_o is expressed in terms of $\{x_k\}$ and β. Although an analytical solution may confound us, there is nothing to stop us tackling the problem numerically. The most straightforward method is to use brute force and ignorance: simply evaluate L, from eqn (2.38), for a whole series of different possible values of α; the number giving rise to the largest L will be the best estimate. This is a tedious procedure if carried out manually, but can be performed painlessly if automated through a short computer program. If we plot the exponential of L, $\exp(L)$, on the vertical axis, as a function of α, on the horizontal axis, then we obtain the posterior pdf for the position of the lighthouse. Not only does this give us a complete visual representation of our inference, it has the advantage that we don't need to worry whether the posterior pdf is asymmetric or multimodal.

Before illustrating the use of such an analysis, it is worth making a brief practical point: it is better to calculate L first, as a function of α (on a uniform grid), and then take exponentials, rather than working directly with the posterior pdf itself. The reason is purely numerical and stems from the limit on the size of the number which can be stored in a computer; typically, anything smaller than 10^{-300} tends to be indistinguishable from zero and numbers larger than 10^{+300} are effectively infinite. The reader may be forgiven for failing to see why this should be a problem, since 600 orders of magnitude is an enormous range; after all, probabilities which are more than a factor of 1000 down (10^{-3}) on the maximum are almost imperceptible on a graph and of little real interest anyway. This is certainly true, but it assumes that the maximum value of the posterior probability has already been scaled to a sensible number. In practice, this is often not the case as we usually ignore the normalization constant until the end of the calculation. In terms of eqn (2.38), we tend to omit the constant term from the sum because it affects neither the best estimate nor its error-bar; its value does, however, translate into a multiplicative scale factor for the posterior pdf. A good procedure, therefore, is to evaluate L for a series of different α's, find the maximum $L_{\max}$ and subtract it from all the L's; taking the exponential of these numbers will give a sensible maximum value of one. As well as being useful for plotting purposes, it then enables us to calculate the normalization constant safely, if required, without fear of underflows or overflows in the execution of the program.

Figure 2.9 shows how the posterior pdf for the position of the lighthouse evolves as the number of flashes detected on the coast increases. A random number generator was used to create a uniform sample of azimuths $\{\theta_k\}$, according to eqn (2.32), which were converted into data positions $\{x_k\}$ with eqn (2.33); the lighthouse was taken to be 1 km out at sea ($\beta = 1$), and assumed to be known. In the early panels, the positions of the flashes are marked by small open circles at the top of the graphs; the number of data analysed is shown in the top right-hand corner. Not only is the posterior pdf very broad

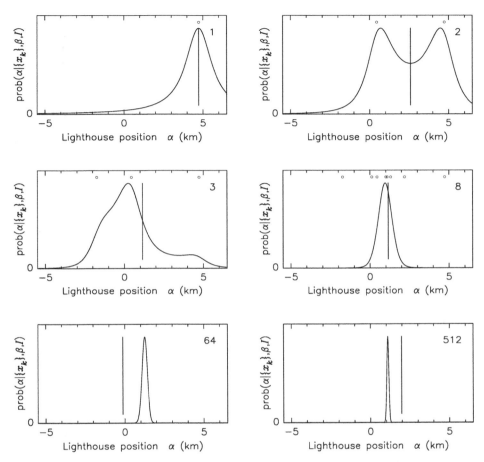

Fig. 2.9 The posterior pdf for the position of the lighthouse along the coast, given that it is 1 km out to sea. The number of data available is shown in the top right-hand corner of the graphs, and their positions are marked by small open circles at the top of the first four panels. The average value of the data is indicated by a long vertical line.

when there are only a few data, it can easily be multimodal if their locations are sufficiently discordant. By about a dozen measurements, however, the posterior becomes a well behaved Gaussian-like pdf. As the experimental evidence mounts, the posterior pdf becomes narrower, and narrower (inversely proportional to $\sqrt{N}$), with the peak converging towards a value of $\alpha = 1$; this is as expected, and quite reassuring, as the data were generated with the lighthouse 1 km along the shore.

2.4.1 The central limit theorem

Before leaving this problem, it is worth pointing out a curious feature about the mean value of the data. In the previous section, we found that the best estimate of the parameter μ of a normal distribution, as in eqn (2.23), was given by the simple arithmetic average of samples drawn from it; stated formally in eqn (2.28), this result seems intuitively reasonable as μ is the mean of the Gaussian pdf. In the present case, the data are drawn from a Cauchy pdf; since this too is symmetric about the point of interest α, we might think that the average value of the data would also provide a good estimate of the position of the lighthouse. We should be a little concerned by this proposal, however, as the sample mean is not the solution to eqn (2.39) for the best estimate α_o. Our suspicions are confirmed in Fig. 2.9, where the average value of the data is indicated by the long vertical line. We see that the sample mean is not a good *estimator* for this problem, since it often lies well outside the range allowed by the posterior pdf.

At first sight, the observation above seems very peculiar; it appears to conflict with our training if not experience. The everyday respectability afforded to averages hinges, formally, on the central limit theorem: this states that if samples are drawn randomly from (almost) any pdf with mean μ, then in the limit of many data N, their average will tend to this value; the error-bar for the difference between μ and the sample mean will also go down like $\sqrt{N}$ (as in eqn 2.29). Technically, it can be shown that the Cauchy distribution is an exception to this rule because it violates one of the underlying assumptions of the central limit theorem. This follows from the fact that means μ and variances σ^2 are additive. On dividing by N to recover averages, the error-bar on the mean diminishes like $\sqrt{N}$. But for this to hold, μ and σ have to exist. For a Cauchy distribution, its very wide wings make σ^2 infinite and μ undefined. In fact, the variability of the mean of the data does not decrease with increasing number of measurements, and is likely to be as 'wrong' after a thousand, or a million, data as it is after just one!

Although the central limit theorem is not applicable to the Cauchy likelihood, this did not prevent us from being able to estimate the position of the lighthouse. To infer the value of α, given β and the detection of flashes at $\{x_k\}$, we needed to consider the posterior pdf $\mathrm{prob}(\alpha | \{x_k\}, \beta, I)$; as can be seen from Fig. 2.9, the latter continues to get sharper as the number of data increases. The point is that the central limit theorem is a statement about the pdf for the sample mean: $\mathrm{prob}(\overline{x} | \alpha, \beta, I)$, where $\overline{x} = \sum x_k / N$. If the maximum of the posterior pdf happens to be equal to the average value of the data, then this might be relevant; otherwise, it's of little concern. The moral of this tale is that, despite its everyday respectability, the sample mean is not always a useful number — let the posterior pdf decide what is best.

3 Parameter estimation II

In the previous chapter, we were concerned with the most elementary type of data analysis problem: parameter estimation involving just one unknown variable. Now let us progress to the more common case in which there are several parameters; some of these will be of interest to us, but others merely a nuisance. As well as being a more challenging optimization problem, we will be led to generalize the idea of error-bars to include correlations and to the use of marginalization to deal with the unwanted variables. We will also see how certain approximations naturally give rise to some of the most frequently used analysis procedures, and discuss the so-called propagation of errors.

3.1 Example 4: amplitude of a signal in the presence of background

In many branches of science, we are often faced with the task of having to estimate the amplitude of a signal in the presence of a background. For example, an X-ray diffraction pattern from a crystalline sample will contain both the distinctive Bragg peaks of interest and a general contribution from diffuse scattering processes; in an astronomical setting, the emission spectrum of a galaxy may be contaminated by stray light from the night sky. An idealized one-dimensional situation is illustrated in Fig. 3.1, where the horizontal x-axis represents the measurement variable (such as the scattering angle, wavelength, or whatever). In the simplest case, the background can be taken as being flat, and of unknown magnitude B, and the signal of interest assumed to be the amplitude A of a peak of known shape and position. The data for these problems are frequently integer-valued, and correspond to things like the number of photons detected from a quasar at a certain wavelength or the number of neutrons scattered by a sample in a given direction. Given such a set of counts $\{N_k\}$, measured at experimental settings $\{x_k\}$, what is our best estimate of the amplitude of the signal peak and the background?

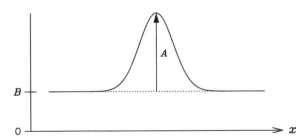

Fig. 3.1 A signal peak of amplitude A sitting on a flat background of magnitude B.

Let us begin by thinking about the nature of the data a little more carefully. It seems reasonable that we should expect the number of counts in the kth data-channel to be proportional to the sum of the signal and the background at x_k; taking the peak to be Gaussian in shape, for the sake of argument, with a width of w and centred at x_o, the ideal datum D_k ought to be given by:

$$D_k = n_o\left[A\,e^{-(x_k-x_o)^2/2w^2} + B\right],\tag{3.1}$$

where n_o is a constant related to the amount of time for which the measurements were made. Unlike the number of counts N_k, however, the D_k in eqn (3.1) will not generally be a whole number. Therefore, the actual datum will be an integer ($\geqslant 0$) in the vicinity of this expected value. A pdf which incorporates this property, and is usually invoked in such counting experiments, is a Poisson distribution:

$$\mathrm{prob}(N|D) = \frac{D^N e^{-D}}{N!}\,,\tag{3.2}$$

and is illustrated in Fig. 3.2. We leave the derivation of this strange-looking formula to Chapter 5 but note the result that its formal expectation value, as defined by the discrete version of eqn (2.22), is indeed equal to D:

$$\langle N\rangle = \sum_{N=0}^{\infty} N\,\mathrm{prob}(N|D) = D\,.\tag{3.3}$$

Equation (3.2) is, of course, an assignment for the likelihood of the datum N_k:

$$\mathrm{prob}(N_k|A,B,I) = \frac{D_k^{N_k}\,e^{-D_k}}{N_k!}\,,\tag{3.4}$$

where the background information I includes a knowledge of the relationship between the expected number of counts D_k and the parameters of interest A and B; for the case of the Gaussian peak-shape model of eqn (3.1), this means that x_o, w and n_o are taken as given (as well as x_k). If the data are independent, so that, when given the values of

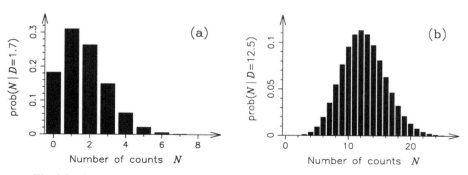

Fig. 3.2 The Poisson distribution $\mathrm{prob}(N|D)$ for: (a) $D=1.7$, and (b) $D=12.5$.

A and B, the number of counts observed in one channel does not influence what we would expect to find in another, then the likelihood function is just the product of the probabilities for the individual measurements:

$$\text{prob}(\{N_k\}|A,B,I) = \prod_{k=1}^{M} \text{prob}(N_k|A,B,I) , \qquad (3.5)$$

where there are M data.

Our inference about the amplitude of the signal and the background is embodied in the posterior pdf $\text{prob}(A,B|\{N_k\},I)$; as usual, we use Bayes' theorem to help us calculate it:

$$\text{prob}(A,B|\{N_k\},I) \propto \text{prob}(\{N_k\}|A,B,I) \times \text{prob}(A,B|I) . \qquad (3.6)$$

Having already dealt with the likelihood function above, all that remains is the assignment of the prior pdf. Irrespective of the data, the one thing we do know is that the amplitude of neither the signal nor the background can be negative; the most naïve way of encoding this is through a uniform pdf in the positive quadrant:

$$\text{prob}(A,B|I) = \begin{cases} \text{constant} & \text{for } A \geqslant 0 \text{ and } B \geqslant 0, \\ 0 & \text{otherwise}. \end{cases} \qquad (3.7)$$

Later, in Chapter 5, we will see that this simple flat pdf is not necessarily the one that represents the most prior ignorance for this problem, but it is quite adequate for our present needs. Although we should formally have upper bounds A_{max} and B_{max}, which determine the normalization constant in eqn (3.7), we will just assume they are sufficiently large as not to impose a cut-off on the posterior pdf; that is to say, in eqn (3.6), $\text{prob}(\{N_k\}|A,B,I)$ will have diminished (gradually) to a negligibly small value before these limits are reached.

Multiplying the Poisson likelihood resulting from eqns (3.4) and (3.5) by the prior of eqn (3.7), according to Bayes' theorem of eqn (3.6), yields the posterior pdf; its logarithm L is given by

$$L = \log_e\left[\text{prob}(A,B|\{N_k\},I)\right] = \text{constant} + \sum_{k=1}^{M} \left[N_k \log_e(D_k) - D_k\right], \quad (3.8)$$

where the constant includes all terms not involving A and B, and the latter are both positive. Our best estimate of the amplitude of the signal and the background is given by the values of A and B which maximize L; its reliability is indicated by the width, or sharpness, of the posterior pdf about this optimal point. Let us illustrate the use of this analysis with some computer examples.

Four sets of data and the posterior pdfs which result from them are shown in Fig. 3.3. The data were generated with a Poisson random number generator, in accordance with eqn (3.4), from the simple 'Gaussian peak on flat background model' of eqn (3.1). They

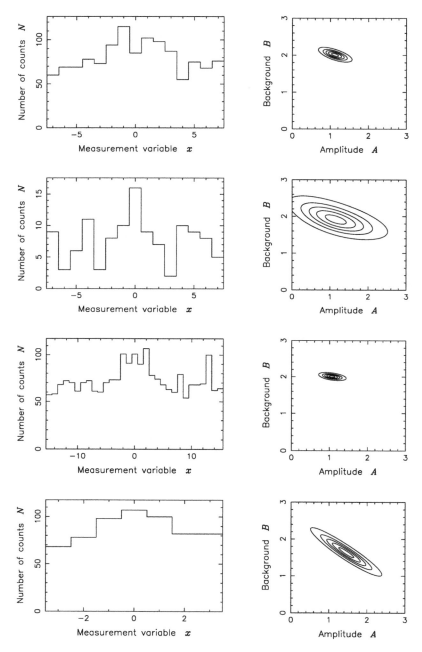

Fig. 3.3 Poisson data and the resulting posterior pdfs for the amplitude A of a Gaussian signal peak, centred at the origin with a FWHM of 5 units, and the flat background B, for four different experimental set-ups.

are plotted as *histograms*, as experimental measurements would normally correspond to the number of counts detected in finite-sized channels; for simplicity, we have chosen the bin-width to be unity. In all cases, the underlying signal is centred at the origin, so that $x_0 = 0$, and has a FWHM of 5 units; this is assumed to be known for the analysis. The posterior pdf is now two-dimensional, since it is a function of both A and B. We can display it using *contours*, which are lines joining points of equal probability density, just like hills and valleys are represented on a topographic map. In Fig. 3.3, the contours correspond to 10%, 30%, 50%, 70% and 90% of the maximum probability.

The first panel shows the number of counts detected in 15 data-bins, where the parameter n_0 was chosen to give a maximum expectation of 100 counts; its value is taken as given. The corresponding posterior pdf, resulting from the exponential of eqn (3.8), is plotted in the second panel on the top; it indicates that our best estimate of the amplitude of the signal is approximately equal to one, and is about half the magnitude of the background. The second panel down shows the data for the same set-up, but where the experiment was conducted for only one-tenth of the time; the number of counts is down by a factor of ten, and the data appear more noisy. The corresponding posterior pdf is about three times as broad as before, in both directions, and is in line with the $\sqrt{10}$ that we might have guessed, on the basis of our experience from the examples of the previous chapter. The posterior is truncated (or suddenly equal to zero) for negative values of A and reflects the importance of the prior when the data are of poor quality. In the third panel down we return to the original count-rate, but have 31 data spread over twice the measurement range. With the doubling of the number of data, we might expect the reliability with which A and B can be estimated would improve by a factor of $\sqrt{2}$; although a little hard to tell from these diagrams, this only seems to be true for the background. Some further thought reveals this to be reasonable: measurements far away from the origin only tell us about the background, since they contain no information about the signal peak. The last two panels illustrate the case when there are just 7 data spread over half the original range in x; the posterior pdf is noticeably broadened, by more than a simple factor of $\sqrt{2}$, and is distinctly skewed. These features are indicative of a strong correlation between our estimates of A and B: as the range of x over which the data are collected is severely restricted, it becomes difficult to tell the signal apart from the background!

3.1.1 Marginal distributions

The two-dimensional posterior pdf above describes completely our joint inference about the values of A and B. Often, however, we are not interested in the background. It has to be included in our calculation, because it is needed for the likelihood function, but we have no intrinsic interest in knowing its value: its presence is merely a nuisance. What we would really like to estimate is just the amplitude of the signal, irrespective of the size of the background; in other words, we require the pdf $\mathrm{prob}(A|\{N_k\}, I)$. According to the marginalization rule of eqn (1.4), this can be obtained simply by integrating the joint posterior pdf with respect to B:

$$\text{prob}(A|\{N_k\}, I) = \int\limits_0^\infty \text{prob}(A, B|\{N_k\}, I)\,\mathrm{d}B\,. \tag{3.9}$$

There may, of course, be situations where the background is of primary interest and the presence of a peak is a nuisance. This seemingly perverse case does sometimes occur, because the phenomena that give rise to broad features in the data may be of as much scientific importance as the ones resulting in a sharp signal. All we need to do then, to obtain the pdf $\text{prob}(B|\{N_k\}, I)$, is integrate the joint posterior pdf with respect to A:

$$\text{prob}(B|\{N_k\}, I) = \int\limits_0^\infty \text{prob}(A, B|\{N_k\}, I)\,\mathrm{d}A\,. \tag{3.10}$$

The four pairs of marginal distributions corresponding to the data-sets, and posterior pdfs, of Fig. 3.3 are plotted in Fig. 3.4; it is now much easier to see the effects of the different experimental set-ups on the reliability of the inferred values A and B, which were mentioned a little earlier.

To avoid any confusion, we should emphasize that the marginal distribution for A, $\text{prob}(A|\{N_k\}, I)$, is not the same as the conditional pdf $\text{prob}(A|\{N_k\}, B, I)$; although both describe our inference about the amplitude of the signal peak, they correspond to different circumstances. The former pdf takes into account our prior ignorance of the value of B, while the latter is appropriate when the magnitude of the background has already been determined reliably from a calibration experiment. This conditional pdf, given that $B = 2$, is plotted as a dotted line in Fig. 3.4; compared with its marginal counterpart, we see that there is a significant narrowing of the posterior probability for the last data-set, and least difference for the case when measurements have been made far beyond the tails of the signal peak. When we are able to distinguish between the background and the signal peak fairly well, there is little to be gained from a separate calibration experiment for B; if the data-set is severely truncated, or the background is highly structured, this additional information can be very beneficial.

For the above analysis, we have assumed that the shape and position of the signal peak is known; in the context of the Gaussian model of eqn (3.1), we took the values of w and x_0 as implicitly given in the background information I. How could this condition be relaxed if we were not quite so fortunate? Well, from the discussion about marginalization, it follows that we should integrate out the relevant variables as nuisance parameters:

$$\text{prob}(A, B|\{N_k\}, I) = \iint \text{prob}(A, B, w, x_0|\{N_k\}, I)\,\mathrm{d}w\,\mathrm{d}x_0\,, \tag{3.11}$$

where I still assumes that the model of eqn (3.1) is adequate, but does not necessarily include a knowledge of the width and position of the Gaussian peak. The four-parameter posterior pdf under the double integral can itself be related to two others which are easier to assign by using Bayes' theorem:

$$\text{prob}(A, B, w, x_0|\{N_k\}, I) \propto \text{prob}(\{N_k\}|A, B, w, x_0, I) \times \text{prob}(A, B, w, x_0|I)\,.$$

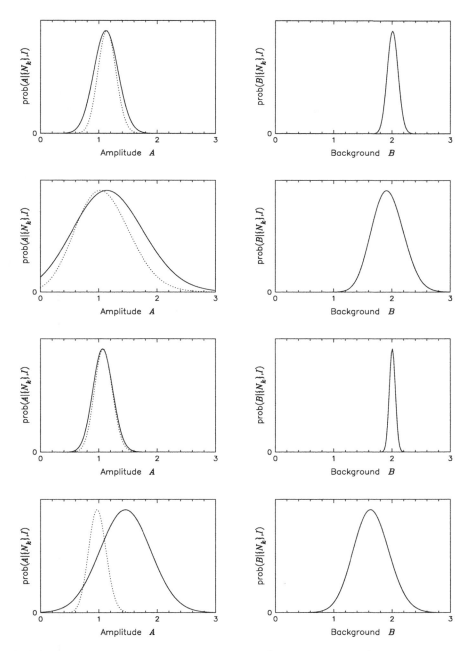

Fig. 3.4 The marginal distributions for the amplitude A and the background B corresponding to the Poisson data, and the resulting posterior pdfs, for the four experimental set-ups in Fig. 3.3. The dotted line shows the posterior pdf for A conditional on knowing the true value of B.

The first term on the right-hand side is equivalent to the likelihood function of eqns (3.4) and (3.5); the second is the prior pdf for A, B, w and x_o, which can be decomposed as:

$$\text{prob}(A, B, w, x_o | I) = \text{prob}(A, B | I) \times \text{prob}(w, x_o | I) .$$

If we already knew the width and location of the Gaussian peak, then the prior pdf for w and x_o would be very sharp. In the limit of absolute certainty, it would become

$$\text{prob}(w, x_o | I) = \delta(w - 2.12) \, \delta(x_o) ,$$

where the two δ-functions are equal to zero unless $w = 2.12$ (i.e. FWHM = 5) and $x_o = 0$ respectively; in this case, the integral of eqn (3.11) is very easy to evaluate and yields

$$\text{prob}(A, B | \{N_k\}, I) \propto \text{prob}(\{N_k\} | A, B, w = 2.12, x_o = 0, I) \times \text{prob}(A, B | I) .$$

Thus, as expected, the analysis reduces to that of eqn (3.6). If we do not know the values of w and x_o, however, we must assign a suitably broad prior for these parameters (as well as for A and B). The marginalization integral will then require more work, but can either be computed numerically or approximated analytically, as we shall see in Section 3.2. Although probability theory allows us to deal with such experimental uncertainties, we do, of course, expect the marginal distribution $\text{prob}(A, B | \{N_k\}, I)$ to be broader than the conditional pdf $\text{prob}(A, B | \{N_k\}, w, x_o, I)$; this would merely reflect the corresponding lack of relevant information.

3.1.2   Binning the data

When the data were plotted in Fig. 3.3 using histograms, we said that this was because experimental measurements often corresponded to the number of counts detected in channels of finite width. This means that eqn (3.1), for the expected value of the datum D_k, should actually be written as an integral over the kth data-bin:

$$D_k = \int_{x_k - \Delta/2}^{x_k + \Delta/2} n_o \left[A\, e^{-(x - x_o)^2 / 2w^2} + B \right] dx , \qquad (3.12)$$

where we have taken all the measurement channels to have the same width Δ. As long as the bin-width is not too large, so that the integrand varies only linearly across it, the integral of eqn (3.12) can be approximated well by the area of a trapezoidal column:

$$D_k \approx n_o \left[A\, e^{-(x_k - x_o)^2 / 2w^2} + B \right] \Delta . \qquad (3.13)$$

Thus, eqn (3.1) is justified because the fixed number Δ can be absorbed into n_o. This new (redefined) constant reflects both the amount of time for which the experimental measurements were made and the size of the 'collecting area'. The bin-width Δ is not always determined by the physical size of the detectors, however, but is often chosen to be large enough so that there are a reasonable number of counts in the resulting composite data-channels. Is there anything to be gained, or lost, in this binning process?

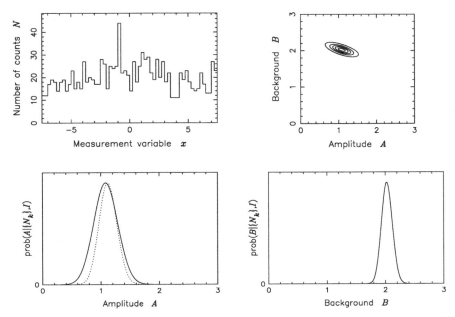

Fig. 3.5 The analysis of data corresponding to the experimental set-up of the first panel in Fig. 3.3, but with bins which are four times narrower.

Figure 3.5 shows the data for the same experimental set-up as in the first panel of Fig. 3.3, but where the bins are four times narrower than before; this data-set looks noisier because, on average, each channel has only a quarter of the previous number of counts. Figure 3.5 also shows the resulting posterior pdf for A and B, and its marginal distributions, given that $w = 2.12$ and $x_0 = 0$. Compared with the corresponding pdfs in Figs 3.3 and 3.4, the reliability of the inferred parameters is virtually identical. This illustrates that the perceived improvement in the data quality afforded by the binning procedure is largely cosmetic. While this visual gain is not without value, and there are computational advantages in dealing with fewer 'measurements', we must always be aware that too coarse a binning can destroy useful information in the data. To take an extreme case, if all the counts were added together into a single number, we would completely lose all our ability to distinguish the signal from the background! At a more mundane level, if the data-channels are too wide, we could also get into trouble with the integral approximation of eqn (3.13). Some of the issues related to optimal binning strategies are discussed further in Chapter 7.

3.2 Reliabilities: best estimates, correlations and error-bars

We have now seen an example of an estimation problem involving more than one parameter. Although the posterior pdf is of a higher dimensionality, being a function of several variables, it still encodes our inference about the values of the parameters, given the data and the relevant background information. As before, we often wish to summarize it with just a few numbers: the best estimates and a measure of their reliabilities. Since the probability (density) associated with any particular set of values for the parameters is a measure of how much we believe that they (really) lie in the neighbourhood of those values, our optimal estimate is given by the maximum of the posterior pdf. If we denote the quantities of interest by $\{X_j\}$, with a posterior pdf $P = \mathrm{prob}(\{X_j\}|\{data\},I)$, then the best estimate of their values, $\{X_{0j}\}$, is given by the solution to the set of *simultaneous equations*:

$$\left.\frac{\partial P}{\partial X_i}\right|_{\{X_{0j}\}} = 0\,, \tag{3.14}$$

where $i = 1, 2, \ldots,$ up to the number of parameters to be inferred. Strictly speaking, we also need a further test, analogous to eqn (2.9), to ensure that we are dealing with a maximum and not a minimum or a *saddle-point*; we'll say more about that shortly. In writing the *partial derivatives* of P with respect to the X_i's we are, of course, assuming that they can all lie in a continuous range. If some could only take discrete values, our best estimate would still correspond to the one with greatest posterior probability but we couldn't then use the calculus notation of eqn (3.14).

As in Section 2.2, it is more convenient to work with the logarithm of P rather than use the posterior pdf itself:

$$L = \log_e\left[\mathrm{prob}(\{X_j\}|\{data\},I)\right]. \tag{3.15}$$

Since the logarithm is a monotonic function, the maximum of L occurs at the same place as that of P; hence, eqn (3.14) can be written with L substituted for P. Rather than pursuing this analysis in generality, with several parameters, let us first consider the specific case of just two variables; we will denote them by X and Y, instead of X_1 and X_2, to reduce the multiplicity of subscripts. The pair of simultaneous equations which we must now solve, to obtain our best estimates X_0 and Y_0, are given by

$$\left.\frac{\partial L}{\partial X}\right|_{X_0,Y_0} = 0 \quad \text{and} \quad \left.\frac{\partial L}{\partial Y}\right|_{X_0,Y_0} = 0\,, \tag{3.16}$$

where $L = \log_e\left[\mathrm{prob}(X,Y|\{data\},I)\right]$.

To obtain a measure of the reliability of this best estimate, we need to look at the spread of the two-dimensional posterior pdf about the point (X_0,Y_0). As in Section 2.2, we can analyse the local behaviour of a (potentially) complicated function by using a Taylor series expansion:

$$L = L(X_0,Y_0) + \frac{1}{2}\left[\left.\frac{\partial^2 L}{\partial X^2}\right|_{X_0,Y_0}(X-X_0)^2 + \left.\frac{\partial^2 L}{\partial Y^2}\right|_{X_0,Y_0}(Y-Y_0)^2\right.$$
$$\left. + 2\left.\frac{\partial^2 L}{\partial X\,\partial Y}\right|_{X_0,Y_0}(X-X_0)\,(Y-Y_0)\right] + \cdots\,, \tag{3.17}$$

where X_0 and Y_0 are given by the condition of eqn (3.16). Although this expression looks quite horrendous, it is simply the two-dimensional version of eqn (2.11); it's just that there are now four second (partial) derivatives to deal with, instead of only one! We have, in fact, reduced this tally of terms to three by using the equality of the mixed derivatives: $\partial^2 L/\partial X \, \partial Y = \partial^2 L/\partial Y \, \partial X$. The first term in the Taylor series, $L(X_O, Y_O)$, is a constant and tells us nothing about the shape of the posterior pdf. The two linear terms, which would be proportional to $X - X_O$ and $Y - Y_O$, are missing because we are expanding about the maximum (as indicated by eqn 3.16). The three quadratic terms are, therefore, the dominant factors determining the width of the posterior pdf, and play a central rôle in the reliability analysis; let us study them more closely.

To aid the generalization of this discussion to the case of several variables a little later, let us rewrite the quadratic part of eqn (3.17) in *matrix* notation; calling the quantity in the square brackets Q, we have

$$Q = \begin{pmatrix} X - X_O & Y - Y_O \end{pmatrix} \begin{pmatrix} A & C \\ C & B \end{pmatrix} \begin{pmatrix} X - X_O \\ Y - Y_O \end{pmatrix} , \tag{3.18}$$

where the components of the 2×2 *symmetric* matrix in the middle are given by the second derivatives of L, evaluated at the maximum (X_O, Y_O):

$$A = \left. \frac{\partial^2 L}{\partial X^2} \right|_{X_O, Y_O} , \quad B = \left. \frac{\partial^2 L}{\partial Y^2} \right|_{X_O, Y_O} , \quad C = \left. \frac{\partial^2 L}{\partial X \, \partial Y} \right|_{X_O, Y_O} . \tag{3.19}$$

Figure 3.6 shows a contour of Q in the X–Y plane; within our quadratic approximation, it is also a line along which the posterior pdf is constant. It is an *ellipse*, centred at (X_O, Y_O), the orientation and *eccentricity* of which are determined by the values of A, B and C; for a given contour-level $(Q = k)$, they also govern its size. The directions of the *principal axes* formally correspond to the *eigenvectors* of the second-derivative

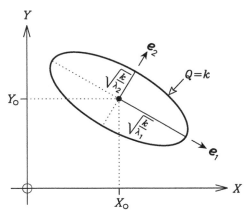

Fig. 3.6 The contour in the X–Y parameter space along which $Q = k$, a constant. It is an ellipse, centred at (X_O, Y_O), the characteristics of which are determined by the eigenvalues λ and eigenvectors e of the second-derivative matrix defined in eqns (3.18)–(3.20).

matrix of eqns (3.18) and (3.19); that is to say, the (x, y) components of e_1 and e_2 in Fig. 3.6 are given by the solutions of the equation

$$\begin{pmatrix} A & C \\ C & B \end{pmatrix} \begin{pmatrix} x \\ y \end{pmatrix} = \lambda \begin{pmatrix} x \\ y \end{pmatrix} . \tag{3.20}$$

The two *eigenvalues* λ_1 and λ_2 which satisfy eqn (3.20) are, in turn, inversely related to the square of the widths of the ellipse along its principal directions. Furthermore, if the point (X_O, Y_O) in eqn (3.16) is to be a maximum rather than a minimum or a saddle-point, both λ_1 and λ_2 must be negative; in terms of A, B and C, this requirement takes the form

$$A < 0 , \qquad B < 0 , \qquad AB > C^2.$$

Returning to the question of the reliability of our best estimates, this would be easy to define if the ellipse in Fig. 3.6 was aligned with the X and Y coordinate axes; this is so when $C = 0$. In that case, the error-bars for X_O and Y_O will just be inversely proportional to the square root of (the modulus of) the appropriate eigenvalues (which are then simply A and B). What should we do if the ellipse is skewed?

If we were only interested in knowing the value of X, then we would integrate out Y as a nuisance parameter:

$$\text{prob}(X | \{data\}, I) = \int\limits_{-\infty}^{+\infty} \text{prob}(X, Y | \{data\}, I) \, dY .$$

This integral can be done analytically within the quadratic approximation of eqns (3.17)–(3.19): $\text{prob}(X, Y | \{data\}, I) = \exp(L) \propto \exp(Q/2)$. Assuming that the joint posterior pdf is not (significantly) truncated by the bounds of the prior, Appendix A shows the result to be

$$\text{prob}(X | \{data\}, I) \propto \exp\left(\frac{1}{2} \left[\frac{AB - C^2}{B} \right] (X - X_O)^2 \right) ,$$

where we have omitted the normalization constant. Comparing this to eqns (2.13) and (2.14), we see that the marginal distribution for X is just a simple one-dimensional Gaussian pdf; our best estimate is still X_O, and its associated error-bar σ_X is given by

$$\sigma_X = \sqrt{\frac{-B}{AB - C^2}} . \tag{3.21}$$

We can obtain an analogous result for Y, yielding a marginal error-bar σ_Y:

$$\sigma_Y = \sqrt{\frac{-A}{AB - C^2}} . \tag{3.22}$$

While the expressions for σ_X and σ_Y above provide a useful measure of the reliability of our best estimates, in general, they paint an incomplete picture. To understand this, we must consider the denominator of eqns (3.21) and (3.22) in a bit more detail. The term $AB - C^2$ is, in fact, the *determinant* of the matrix in eqn (3.20); for such a real symmetric matrix, this is equal to the product of its eigenvalues. Thus if either λ_1 or λ_2 becomes very small, so that the ellipse in Fig. 3.6 is extremely elongated in one of its principal directions, then $AB - C^2 \to 0$; consequently, apart from the special case when $C = 0$, both σ_X and σ_Y will be huge. Although this correctly warns us that neither X nor Y can be inferred reliably on the basis of the current data, it fails to tell us that there could still be some joint aspect of the two which can be determined well, because the posterior pdf might be very sharp in one direction while being extremely broad in the other. To see how this information can be conveyed, let us first look at the concept of error-bars again in a slightly different way.

So far, from eqn (2.14), we have thought of the error-bar as representing the width of a Gaussian pdf: FWHM $\approx 2.35\,\sigma$. Another way to think about it is in terms of the *variance* of the posterior, which also gives a measure of its spread. It is formally defined to be the expectation value of the square of the deviations from the mean; assuming a normalized pdf, this is given by

$$\mathrm{Var}(X) = \left\langle (X-\mu)^2 \right\rangle = \int (X-\mu)^2 \, \mathrm{prob}(X|\{data\}, I)\,\mathrm{d}X\,, \qquad (3.23)$$

where μ is the average value $\langle X \rangle$, as in eqn (2.22). For the one-dimensional normal distribution of eqn (2.14), this integral yields the result

$$\left\langle (X-\mu)^2 \right\rangle = \sigma^2\,. \qquad (3.24)$$

The square root of the variance is called the *standard deviation*, or the *root-mean-square* (r.m.s.) error, of X. This definition of the error-bar can be extended to pdfs of more than one variable. Explicitly, for the two-dimensional case we have been considering,

$$\sigma_X^2 = \left\langle (X-X_0)^2 \right\rangle = \iint (X-X_0)^2 \, \mathrm{prob}(X, Y|\{data\}, I)\,\mathrm{d}X\,\mathrm{d}Y\,, \qquad (3.25)$$

where σ_X is the same as in eqn (3.21) when the double integral is evaluated within the quadratic approximation of eqns (3.17)–(3.19). The corresponding expression for σ_Y is similar, with the X's and Y's swapped around.

The idea of variance can be broadened to consider the simultaneous deviations of both X and Y; this *covariance*, which we will denote as σ_{XY}^2, is given by

$$\sigma_{XY}^2 = \left\langle (X-X_0)(Y-Y_0) \right\rangle$$

$$= \iint (X-X_0)(Y-Y_0)\,\mathrm{prob}(X, Y|\{data\}, I)\,\mathrm{d}X\,\mathrm{d}Y\,, \qquad (3.26)$$

and is a measure of the *correlation* between the inferred parameters. If an over-estimate of one usually leads to a larger than average value for the other, then the difference

$Y - Y_O$ will tend to be positive when $X - X_O$ is positive; if the same is true for under-estimates, so that $Y - Y_O$ is usually negative when $X - X_O$ is as well, the expectation value of the product of the deviations will be positive: the covariance will then be greater than zero. If there is an anti-correlation, so that an over-estimate of one is accompanied by an under-estimation of the other, then the covariance will be negative. When our estimate of one parameter has little, or no, influence on the inferred value of the other, then the magnitude of the covariance will be negligible in comparison to the variance terms; in other words, $|\sigma_{XY}^2| \ll \sqrt{\sigma_X^2\,\sigma_Y^2}$.

When the double integral of eqn (3.26) is evaluated within the quadratic approximation of eqns (3.17)–(3.19), it yields the result

$$\sigma_{XY}^2 \;=\; \frac{C}{AB - C^2}\,. \tag{3.27}$$

In conjunction with eqns (3.21) and (3.22), therefore, we see that both the variance and covariance terms are given by (minus) the elements of the *inverse* of the second-derivative matrix of eqn (3.20):

$$\begin{pmatrix} \sigma_X^2 & \sigma_{XY}^2 \\ \sigma_{XY}^2 & \sigma_Y^2 \end{pmatrix} = \frac{1}{AB - C^2}\begin{pmatrix} -B & C \\ C & -A \end{pmatrix} = -\begin{pmatrix} A & C \\ C & B \end{pmatrix}^{-1}. \tag{3.28}$$

This table of 'error-bar products' is called the *covariance matrix*. When $C = 0$, σ_{XY}^2 also equals zero and means that the inferred values of the parameters are uncorrelated. In that case, the principal directions of the corresponding posterior pdf will be parallel to the coordinates axes; this situation is illustrated in Fig. 3.7(a). As the magnitude of C increases, relative to A and B, the posterior pdf becomes more and more skew and elongated; this reflects the growing strength of the correlation between our estimates

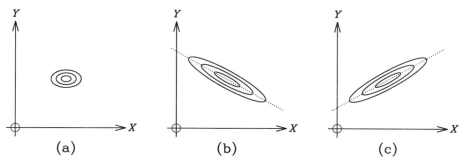

Fig. 3.7 A schematic illustration of covariance and correlation. (a) The contours of a posterior pdf with zero covariance, where the inferred values of X and Y are uncorrelated. (b) The corresponding plot when the covariance is large and negative; $Y + mX = \text{constant}$ along the dotted line (where $m > 0$), emphasizing that only this sum of the two parameters can be inferred reliably. (c) The case of positive correlation, where we learn most about the difference $Y - mX$; this is constant along the dotted line.

of X and Y, and is shown schematically in Figs 3.7(b) and (c). In the extreme case, when $C = \pm\sqrt{AB}$, the elliptical contours will be infinitely wide in one direction (with only the information in the prior preventing this catastrophe!) and oriented at an angle $\pm\tan^{-1}\sqrt{A/B}$ with respect to the X-axis. Although the error-bars σ_X and σ_Y will be huge, saying that our individual estimates of X and Y are completely unreliable, the large off-diagonal elements of the covariance matrix tells us that we can still infer a linear combination of these parameters quite well. If the covariance is negative, then the posterior pdf will be very broad in the direction $Y = -mX$, where $m = \sqrt{A/B}$, but fairly narrow perpendicular to it; this bunching of the probability contours along the lines of '$Y + mX = $ a constant' indicates that the data contain a lot of information about the sum $Y + mX$, but little about the difference $Y - X/m$. Similarly, when the covariance is positive, we can infer the difference $Y - mX$ but not the sum $Y + X/m$.

3.2.1 Generalization of the quadratic approximation

Let us generalize the *bivariate* analysis of the last few pages to the case when there are several variables, and use the opportunity to summarize the main results. The condition for obtaining our best estimate was given in eqn (3.14), and was chosen to be the one which maximized the posterior pdf for the set of M parameters $\{X_j\}$. In terms of its logarithm L, of eqn (3.15), this requirement for the optimal values $\{X_{0j}\}$, or $\boldsymbol{X}_0$ for simplicity, can be expressed by the simultaneous equations

$$\left.\frac{\partial L}{\partial X_i}\right|_{\boldsymbol{X}_0} = 0\,, \tag{3.29}$$

where $i = 1, 2, \ldots, M$. For the *multivariate* case, the extension of the Taylor series expansion of eqn (3.17) takes the form

$$L = L(\boldsymbol{X}_0) + \frac{1}{2}\sum_{i=1}^{M}\sum_{j=1}^{M}\left.\frac{\partial^2 L}{\partial X_i\,\partial X_j}\right|_{\boldsymbol{X}_0}(X_i - X_{0i})(X_j - X_{0j}) + \cdots. \tag{3.30}$$

If the elements of a *column* matrix ($M \times 1$) are regarded as the components of a vector, and we ignore higher-order terms in the Taylor series expansion, then the exponential of eqn (3.30) yields the following approximation for the posterior pdf:

$$\mathrm{prob}(\boldsymbol{X}\,|\,\{data\}, I) \propto \exp\left[\frac{1}{2}(\boldsymbol{X} - \boldsymbol{X}_0)^{\mathrm{T}}\,\boldsymbol{\nabla\nabla}L(\boldsymbol{X}_0)\,(\boldsymbol{X} - \boldsymbol{X}_0)\right], \tag{3.31}$$

where $\boldsymbol{\nabla\nabla}L$ is the symmetric $M \times M$ matrix of second-derivatives, whose ijth element is $\partial^2 L/\partial X_i\,\partial X_j$, and the *transpose* T of $\boldsymbol{X} - \boldsymbol{X}_0$ is a *row* vector. As such, the quadratic exponent is just a generalization of Q in eqn (3.18) and we can think of Fig. 3.6 as being a two-dimensional slice through it. The pdf of eqn (3.31) is called a *multivariate* Gaussian, since it is a multidimensional version of eqns (2.13) and (2.14). Assuming that the bounds of the prior do not cause a significant truncation of this M-dimensional probability 'bump', Appendix A shows the associated constant of proportionality to be given by $(2\pi)^{-M/2}$ times the square root of the determinant of the $\boldsymbol{\nabla\nabla}L$ matrix; this

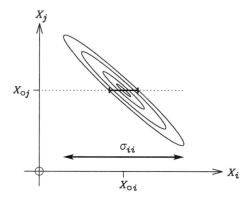

Fig. 3.8 An illustration of the difference between the (correct) marginal error-bar σ_{ii} and its (misleading) 'best-fit' counterpart indicated by the bold line in the middle. The former is given by the ith diagonal element of $(\boldsymbol{\nabla}\boldsymbol{\nabla}L)^{-1}$, while the latter is related to the inverse of $\partial^2 L/\partial X_i^2$.

ensures the correct normalization, so that the multidimensional integral of the posterior pdf with respect to all the parameters $\{X_j\}$ is equal to unity.

It will come as no surprise that the maximum of the multivariate Gaussian is defined by the vector $\boldsymbol{X}_O$; the condition for finding its components, in eqn (3.29), can be written compactly as: $\boldsymbol{\nabla}L(\boldsymbol{X}_O)=0$. By comparison with the standard one-dimensional Gaussian of eqn (2.14), we see that $\boldsymbol{\nabla}\boldsymbol{\nabla}L$ is analogous to $-1/\sigma^2$; this suggests that the spread (or 'width') of the posterior should be related to the inverse of the second-derivative matrix. Indeed, as shown in Appendix A, the covariance matrix σ^2 is given by minus the inverse of $\boldsymbol{\nabla}\boldsymbol{\nabla}L$ (evaluated at $\boldsymbol{X}_O$):

$$\left[\sigma^2\right]_{ij} = \left\langle (X_i - X_{Oi})(X_j - X_{Oj}) \right\rangle = -\left[(\boldsymbol{\nabla}\boldsymbol{\nabla}L)^{-1}\right]_{ij}, \tag{3.32}$$

and is a generalization of eqns (3.26) and (3.28). The square root of the diagonal elements ($i=j$) corresponds to the (marginal) error-bars for the associated parameters; the off-diagonal components ($i \neq j$) tell us about the correlations between the inferred values of X_i and X_j.

Before concluding our discussion of the quadratic approximation, it is worth emphasizing the fact that the inverse of the diagonal elements of a matrix are not, in general, equal to the diagonal elements of its inverse. Stated in this direct way, few of us would expect to make the mistake. We would be pursuing this folly inadvertently, however, if we were foolishly tempted to estimate the reliability of one parameter in a multivariate problem by holding all the others fixed at their optimal values. The situation is illustrated schematically in Fig. 3.8 and shows that the estimate of the error-bars can be misleadingly small if we try to avoid the marginalization procedure.

3.2.2   Asymmetric and multimodal posterior pdfs

The above analysis, leading to the approximation of the posterior pdf by a multivariate Gaussian, relies on the validity of the quadratic expansion of eqn (3.30). The elliptical

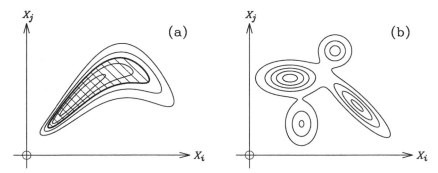

Fig. 3.9 A schematic illustration of two posterior pdfs for which the multivariate Gaussian is not a good approximation: (a) a 'lop-sided' pdf, analogous to the asymmetric case of Section 2.2; (b) a multimodal pdf.

contours in Fig. 3.3 indicate that this is often a reasonable assumption, although we must be careful about their potentially truncated nature when there are few data. There are situations, however, when the posterior can have a very lop-sided shape; such a pdf is sketched in Fig. 3.9(a). The word 'asymmetric' may not be appropriate here, because there might still be a high degree of symmetry involved, but our difficulty is akin to the case discussed earlier in Section 2.2: namely, the concepts of the error-bar and linear correlation associated with the covariance matrix do not seem suitable; at the very least, the correspondence between the integral definition of covariance and the inverse of $\nabla\nabla L$ will be doubtful. We might think of generalizing the idea of confidence intervals to a multidimensional space, but it will usually be hard to describe the surface of the (smallest) hypervolume containing 90% of the probability (say) in just a few numbers. The prescriptions of Section 2.2 can, of course, be used for the one-dimensional marginal distributions $\mathrm{prob}(X_j\,|\,\{data\}, I)$ for each of the parameters; unfortunately, the relevant multiple integrals over the joint pdf will often be difficult to carry out (analytically), and only tell part of the story. Without the convenient properties of the multivariate Gaussian, the maximum of the marginal pdf for X_j may not be the same as the jth component of the vector X_O either; thus we will have to think carefully whether we are interested in the best estimate of all the parameters simultaneously, or just the optimal value of one irrespective of the others.

The situation may be even worse, because the posterior might have many maxima; such a pdf is shown schematically in Fig. 3.9(b). For some types of problems, this multimodal nature can persist no matter how good the quality of the data. There is no difficulty in defining the best estimate, and expressing its reliability, when one of the probability bumps is much bigger than the others: we simply make a (local) quadratic expansion about the global maximum, and ignore the subsidiary solutions. If several of the maxima are of comparable magnitude, then we could carry out this procedure for all of them and provide a list of the alternative (almost) optimal values and their associated covariance matrices; if the number of significant solutions is very large, however, then we could soon be in trouble. Actually, we tend to have serious difficulties with multimodal pdfs even when there is an obvious single best estimate: the problem is the practical one of how (to write a computer program) to find the global solution in a large multidimensional parameter space when there are many (small) local 'hillocks' to trip us up along the way. We will return to this topic in Section 3.4, when discussing some aspects about the computer implementation of our Bayesian analyses. First, though, let us briefly take another look at the Gaussian noise problem in the light of the multivariate character of this chapter.

3.3   Example 5: Gaussian noise revisited

In Section 2.3, we were concerned with estimating the mean μ of a Gaussian process. Since that chapter was restricted to a consideration of one-parameter examples, we assumed that the magnitude of the error-bar σ for the experimental measurements $\{x_k\}$ was already known. Now we are in a position to be able to relax that condition. What we require is the pdf $\text{prob}(\mu|\{x_k\}, I)$, rather than the corresponding posterior conditional on being given the variance $\text{prob}(\mu|\{x_k\}, \sigma, I)$. As a Gaussian process is defined by both μ and σ, we must integrate out the standard deviation as a nuisance parameter from their joint posterior pdf:

$$\text{prob}(\mu|\{x_k\}, I) = \int_0^\infty \text{prob}(\mu, \sigma|\{x_k\}, I)\, d\sigma\,, \tag{3.33}$$

where the integrand can be expressed as a product of the likelihood function and the prior, for the two parameters, by using Bayes' theorem:

$$\text{prob}(\mu, \sigma|\{x_k\}, I) \propto \text{prob}(\{x_k\}|\mu, \sigma, I) \times \text{prob}(\mu, \sigma|I)\,. \tag{3.34}$$

As in eqn (2.25), if the data are independent, the likelihood function is simply the product of the probabilities for obtaining the individual measurements; using the definition of a Gaussian distribution stated in eqn (2.23), we obtain

$$\text{prob}(\{x_k\}|\mu, \sigma, I) = \left(\sigma\sqrt{2\pi}\right)^{-N} \exp\left[-\frac{1}{2\,\sigma^2}\sum_{k=1}^N (x_k-\mu)^2\right], \tag{3.35}$$

where there are N data.

What about the pdf $\text{prob}(\mu, \sigma|I)$? To express complete prior ignorance about these parameters we should assign a pdf that is uniform with respect to μ and $\log\sigma$; the reason for this peculiar choice will become apparent in Chapter 5, but stems from the fact that the position μ is associated with an additive uncertainty whereas the width σ is a multiplicative scale-factor. For the moment, however, let us pursue our policy of *naïveté* and use a simple flat prior:

$$\text{prob}(\mu, \sigma|I) = \begin{cases} \text{constant} & \text{for } \sigma > 0\,, \\ 0 & \text{otherwise}\,. \end{cases} \tag{3.36}$$

We will indicate how the results would be different if we had made the preferred assignment mentioned above, but our conclusions will be unaffected for all practical purposes.

Multiplying the prior of eqn (3.36) by the likelihood function of eqn (3.35), as in eqn (3.34), allows us to write the marginal distribution of eqn (3.33) as

$$\text{prob}(\mu|\{x_k\}, I) \propto \int_0^\infty t^{N-2} \exp\left[-\frac{t^2}{2} \sum_{k=1}^N (x_k - \mu)^2\right] \mathrm{d}t , \qquad (3.37)$$

where we have made the substitution $\sigma = 1/t$ (so that $\mathrm{d}\sigma = -\mathrm{d}t/t^2$). A rescaling of t through the substitution $\tau = t\sqrt{\Sigma (x_k - \mu)^2}$ then reduces eqn (3.37) to

$$\text{prob}(\mu|\{x_k\}, I) \propto \left[\sum_{k=1}^N (x_k - \mu)^2\right]^{-(N-1)/2} , \qquad (3.38)$$

because a definite integral involving only τ can be absorbed into the proportionality. The analysis of Section 2.2 shows that we can obtain our best estimate μ_o and a measure of its reliability from the first and second derivatives of the logarithm of this (marginal) posterior pdf. Accordingly,

$$\left.\frac{\mathrm{d}L}{\mathrm{d}\mu}\right|_{\mu_\mathrm{o}} = \frac{(N-1)\sum (x_k - \mu_\mathrm{o})}{\sum (x_k - \mu_\mathrm{o})^2} = 0 ,$$

where $L = \log_\mathrm{e}[\text{prob}(\mu|\{x_k\}, I)]$ and the summations are from $k = 1$ to N; this can only be satisfied if the numerator is equal to zero, and yields the result of eqn (2.28):

$$\mu_\mathrm{o} = \frac{1}{N} \sum_{k=1}^N x_k . \qquad (3.39)$$

In other words, the optimal estimate of μ is still given by the arithmetic average of the data. Differentiating L for a second time, and evaluating it at the maximum, we have

$$\left.\frac{\mathrm{d}^2 L}{\mathrm{d}\mu^2}\right|_{\mu_\mathrm{o}} = -\frac{N(N-1)}{\sum (x_k - \mu_\mathrm{o})^2} .$$

Since eqn (2.15) tells us that the error-bar for the best estimate is given by the inverse of the square root of (minus) the second derivative, we can summarize our inference of the mean by

$$\mu = \mu_\mathrm{o} \pm \frac{S}{\sqrt{N}} , \quad \text{where } S^2 = \frac{1}{N-1} \sum_{k=1}^N (x_k - \mu)^2 . \qquad (3.40)$$

Comparing this to eqn (2.29), we see that the only difference with the case of Section 2.3 is that the formerly (assumed) known value of σ has been replaced by an estimate derived from the data.

3.3.1 The Student-t and χ^2 distributions

In Section 2.3, we noted that the posterior pdf $\mathrm{prob}(\mu|\{x_k\}, \sigma, I)$ was defined completely by the best estimate and its associated error-bar; this was because the quadratic termination of the Taylor series in eqn (2.11) was exact for that case. When σ is not known beforehand, this is no longer true: the summary of eqn (3.40) just represents a useful approximation to the (marginal) posterior $\mathrm{prob}(\mu|\{x_k\}, I)$. The actual pdf is given in eqn (3.38); its shape is easier to ascertain if we rewrite the sum $\sum(x_k - \mu)^2$ in terms of two parameters, $\bar{x}$ and V, derived from the data:

$$\sum_{k=1}^{N} (x_k - \mu)^2 \;=\; N\,(\bar{x} - \mu)^2 + V\,, \tag{3.41}$$

where $\bar{x}$ is equal to the sample-mean of eqn (3.39) and V is given by

$$V \;=\; \sum_{k=1}^{N} (x_k - \bar{x})^2\,. \tag{3.42}$$

Substituting for the sum from eqn (3.41) into eqn (3.38) yields

$$\mathrm{prob}(\mu|\{x_k\}, I) \;\propto\; \left[N\,(\bar{x} - \mu)^2 + V \right]^{-(N-1)/2}\,, \tag{3.43}$$

and is called a *Student-t* distribution. When $N = 3$, this pdf has the same form as the Cauchy distribution of eqn (2.34) which was plotted in Fig. 2.8. It has a maximum at $\mu = \bar{x}$, a FWHM which is proportional to $\sqrt{V}$ and very long tails. As the number of data increases, this type of function is multiplied by itself many times over; as a result, the wide wings of the pdf are soon killed off to leave a more Gaussian-like distribution centred about $\bar{x}$. Hence, the optimal value μ_{o} is always equal to the sample-average; its error-bar, which is related to V through eqn (3.40), becomes more meaningful as N gets larger ($\geqslant 10$).

   In our short discussion about the prior earlier, we promised to indicate how the results would differ if we had assigned a pdf which was uniform with respect to $\log \sigma$. We would have been led to the posterior pdf

$$\mathrm{prob}(\mu|\{x_k\}, I) \;\propto\; \left[N\,(\bar{x} - \mu)^2 + V \right]^{-N/2}\,, \tag{3.44}$$

which is identical to eqn (3.43) except that the power of the $N/2$-term has increased by one; that is, we would have obtained a Student-t distribution with $N-1$ *degrees-of-freedom*, instead of $N-2$. Their shapes are very similar, with a maximum at $\bar{x}$, but the pdf of eqn (3.44) is a little narrower than that of eqn (3.43). In terms of eqn (3.40), S^2 would be given by V/N rather than $V/(N-1)$; this difference is negligible if N is moderately large, which is also the régime when the concept of an error-bar is most appropriate. Thus the simple-minded assignment of eqn (3.36) gives us a slightly more conservative estimate of the reliability of the optimal value μ_{o}, when there are few data, but our conclusions remain essentially unchanged.

Before leaving this section, let us consider what we can learn from the data about the magnitude of the expected error in the measurements. Our inference about σ is described by the (marginal) posterior pdf $\text{prob}(\sigma \mid \{data\}, I)$:

$$\text{prob}(\sigma \mid \{x_k\}, I) = \int\limits_{-\infty}^{+\infty} \text{prob}(\mu, \sigma \mid \{x_k\}, I) \, d\mu \,. \qquad (3.45)$$

Using the likelihood function and prior of eqns (3.35) and (3.36), along with the substitution of eqn (3.41), Bayes' theorem allows to write this as

$$\text{prob}(\sigma \mid \{x_k\}, I) \propto \sigma^{-N} \exp\left(-\frac{V}{2\sigma^2}\right) \int\limits_{-\infty}^{+\infty} \exp\left[-\frac{N(\overline{x} - \mu)^2}{2\sigma^2}\right] d\mu \,, \qquad (3.46)$$

where we have taken all the terms not involving μ outside the integral, and $\sigma > 0$. Since the area under the one-dimensional normal distribution on the right-hand side is proportional to σ, we obtain the result

$$\text{prob}(\sigma \mid \{x_k\}, I) \propto \sigma^{1-N} \exp\left(-\frac{V}{2\sigma^2}\right). \qquad (3.47)$$

In the formal language of statistics, this is related to the χ^2 distribution (through the substitution $X = V/\sigma^2$). As usual, according to the analysis of Section 2.2, the best estimate and its error-bar can be derived from the first and second derivatives of the logarithm of eqn (3.47). This allows us to summarize our inference about σ by

$$\sigma = \sigma_o \pm \frac{\sigma_o}{\sqrt{2(N-1)}} \,, \qquad (3.48)$$

where the optimal value is $\sigma_o = \sqrt{V/(N-1)}$. Although eqn (3.48) permits negative values of σ, this simply indicates that the quadratic expansion of eqn (2.11) gives a poor approximation to the asymmetric posterior pdf when the number of data is small; it would then be better to give an estimate and error-bar for $\log \sigma$ instead, which we will learn how to do in Section 3.6.

3.4 Algorithms: a numerical interlude

We have now seen several examples of how the rules of probability theory can be used to obtain the posterior pdf for the quantities of interest; in order to summarize the inference, we usually need to find its maximum. Sometimes this optimization can be done analytically, but often we are forced to turn to the computer. The virtue of this electronic machine lies, of course, in its ability to do numerical calculations (and graphical plotting) considerably more quickly, and painlessly, than we could do ourselves by hand. Since computer programming and optimization are large subjects in their own right, our aim here is merely to give a flavour of the range of problems that are likely to be encountered and to give an indication of the sort of procedures that might be used to tackle them. A more advanced account of a novel *Monte Carlo* method for doing Bayesian computation is given in Chapters 9 and 10.

We should point out that the material in this section is to do with *algorithms*, rather than the fundamentals of data analysis. That is to say, the techniques employed to find (as opposed to define) the optimal solution are based purely on practical considerations; the fact that we are using them to solve probabilistic problems is essentially coincidental. Details of the procedures mentioned below can be found in numerous good books; of particular note is *Numerical Recipes*, by Press *et al.* (1986), which includes source-code for computer programs (in FORTRAN, C and Pascal). *Practical Optimization*, by Gill *et al.* (1981), is also excellent, as is the old gem by Acton (1970) entitled *Numerical Methods That (usually) Work*.

3.4.1 Brute force and ignorance

For a one-parameter problem, the most robust way of finding the maximum of the posterior pdf is to plot it out and have a look! On a computer, this brute force and ignorance method entails the division of the horizontal axis into a finite number of grid points, representing the possible values of the parameter, and the evaluation of the posterior probability at each of them. Plotting the latter along the vertical axis, we simply find the biggest value and read off the corresponding optimal estimate of the quantity of interest. As mentioned in Section 2.4, it is numerically better to calculate the logarithm of the posterior probability for all the grid points first; and then subtract the largest number from each and exponentiate. If normalization is required, this pdf can be multiplied by a constant so that the area under the curve is equal to unity.

The greatest advantage of this elementary algorithm is that it gives a complete picture of our inference about the value of the parameter. As long as there are enough grid points (a few hundred usually being quite adequate), which cover a sufficiently large range, this procedure will always work. It doesn't matter whether the posterior pdf is asymmetric, multimodal or whatever; even the question of differentiability is of no concern. This brute force method generalizes readily to two-parameter problems: we just plot the posterior probability vertically against a two-dimensional grid of points, often with contours. This is what was done in Fig. 3.3. As well as all the advantages mentioned above, it is very easy to obtain the marginal distributions: simply add up the probabilities in the 'X' or 'Y' directions, as appropriate. Even sharp cut-offs imposed by the prior pdf cause no difficulty for this numerical integration; this can be seen from the low-count example in Figs 3.3 and 3.4.

Unfortunately, this direct approach, of evaluating the probability everywhere, rapidly becomes impractical after the two-parameter case. Apart from the problem of how to display a function of several variables on a piece of paper, the computational cost of calculating the probabilities is soon prohibitive. If each axis was divided into just 10 discrete points, an M-parameter problem would then require 10^M evaluations. While the calculations for two variables might only take a second, a whole day will be required for seven and the age of the universe for twenty! Super-computers, and ridiculously coarse grids, can buy a little more leeway, but not much. For multivariate analysis, we must look for more efficient algorithms.

3.4.2 The joys of linearity

In Section 3.2, we represented a set of M parameters $\{X_j\}$ by the components of a (column) vector X; the condition for our best estimate of their values, denoted by X_O, can be written as

$$\nabla L(X_O) = 0 \,, \tag{3.49}$$

where the jth element of ∇L is given by the partial derivative of the logarithm of the posterior pdf $\partial L/\partial X_j$ (evaluated at $X = X_O$). Equation (3.49) is very compact notation for a set of M simultaneous equations; they will be quite difficult to solve, in general, unless they are *linear*. That is to say, if we are fortunate enough to be able to rearrange ∇L into a form resembling a straight line ('$y = mx + c$'),

$$\nabla L = H X + C \,, \tag{3.50}$$

where the components of the vector C, and the square matrix H, are all constant, then the solution to eqn (3.49) is trivial:

$$X_O = - H^{-1} C \,. \tag{3.51}$$

Differentiating eqn (3.50), we find that the important $\nabla\nabla L$ matrix is invariant with respect to the parameters $\{X_j\}$:

$$\nabla\nabla L = H \,, \tag{3.52}$$

and so all the higher derivatives are identically zero. The covariance matrix σ^2, given by minus the inverse of $\nabla\nabla L$, therefore provides a complete description of the shape of the posterior pdf:

$$\left[\sigma^2\right]_{ij} = \left\langle (X_i - X_{Oi})(X_j - X_{Oj}) \right\rangle = - \left[H^{-1} \right]_{ij} \,,$$

where the $\{X_{Oj}\}$ are the components of the vector X_O in eqn (3.51).

   Although it is easy to write down the optimal solution in the form of eqn (3.51), actually evaluating it can still entail a fair amount of computational effort. Luckily, however, there are plenty of good *procedures*, or *subroutines*, available for matrix algebra; algorithmic details can be found in the references mentioned earlier. If we solved the set of simultaneous equations $H X_O = - C$ directly, with a *Cholesky decomposition* for example, it would avoid the need to invert the matrix H; this should also be somewhat faster numerically for obtaining the best estimate X_O. Whichever route we decide to follow, it will be most efficient if the algorithm used exploits the known symmetry of $\nabla\nabla L$ ($\partial^2 L / \partial X_i \, \partial X_j = \partial^2 L / \partial X_j \, \partial X_i$). The computational time taken for such matrix manipulations tends to scale like the cube of the number of variables M^3; the speed of personal computers is sufficient to tackle most reasonable parameter estimation problems in this way.

   We will, of course, have difficulty in solving the equations for the optimal estimate if the determinant of H is zero, or extremely small; in that case, our answers will be highly sensitive to small changes in the data and the corresponding (marginal) error-bars will be huge. In terms of Fig. 3.6, this will happen if the ellipsoid of $Q = k$, where $Q = (X-X_O)^{\mathrm{T}} \nabla\nabla L(X_O) (X-X_O)$, is (almost) infinitely long in any of its principal directions. In this situation, it is useful to analyse the eigenvalues and eigenvectors of **H**: the principal axes indicate which linear combinations of the parameters can be inferred independently of each other and the eigenvalues give the reliability with which they can be estimated. The only real cure for the predicament is to improve the characteristics of the posterior pdf; this can be done by obtaining more relevant data, or by supplementing them with cogent prior information.

3.4.3 Iterative linearization

The linear problem is so convenient, both analytically and computationally, that it is worth trying to make use of it even when ∇L cannot quite be written in the form of eqn (3.50). To see how we might do this, consider the Taylor series expansion of L about some arbitrary point in parameter space X_1:

$$L = L(X_1) + (X-X_1)^{\mathrm{T}} \nabla L(X_1) + \frac{1}{2}(X-X_1)^{\mathrm{T}} \nabla\nabla L(X_1) (X-X_1) + \cdots .$$

In the past, the first-derivative term has been missing because we've always expanded about the optimal estimate X_O. Differentiating this with respect to the $\{X_j\}$, we obtain

$$\nabla L = \nabla L(X_1) + \nabla\nabla L(X_1) (X-X_1) + \cdots , \qquad (3.53)$$

which is just the Taylor series for ∇L. If we ignore the higher-order terms on the right-hand side, so that eqn (3.53) can be rearranged into the linear form of eqn (3.50), then the solution to eqn (3.49) is given by

$$X_O \approx X_1 - \left[\nabla\nabla L(X_1) \right]^{-1} \nabla L(X_1) . \qquad (3.54)$$

This relationship will be exact when $X_1 = X_O$, or if ∇L is truly linear, but it will be a reasonable approximation as long as X_1 is close enough to the optimal estimate. As such, it lends itself to suggest an *iterative* algorithm: (i) start with a good guess of the optimal solution X_1; (ii) evaluate the gradient-vector ∇L, and the second-derivative matrix $\nabla\nabla L$, at $X = X_1$; (iii) calculate an improved estimate X_2, by equating it to the right-hand side of eqn (3.54); (iv) repeat this process until $\nabla L = 0$.

The procedure outlined above is called the *Newton–Raphson* algorithm. It is a generalization of the numerical technique often used to find the *roots* of a function $f(x_O)=0$; in this case, the function is multivariate: $\nabla L(X_O) = 0$. We can summarize it with the recursion relationship

$$X_{N+1} = X_N - \left[\nabla\nabla L(X_N) \right]^{-1} \nabla L(X_N) , \qquad (3.55)$$

where X_N is our estimate of the solution after $N-1$ iterations. If ∇L is really linear, then only one iteration will be required: X_2 will equal X_O, irrespective of the initial guess X_1. In general, the algorithm will rapidly converge to X_O as long as the starting point is 'close enough' to the optimal solution; we will say more about this shortly, but first let us briefly mention an important practical point.

The stability of iterative procedures can usually be improved by slowing them down a little. This means that it can be advantageous to make a slightly smaller change in going from X_N to X_{N+1} than that recommended by eqn (3.55). Although this can easily be done by multiplying the matrix-vector product on the right of eqn (3.55) by a fractional constant, it tends to be better to achieve a similar effect by adding a small (negative) number c to all the diagonal elements of $\nabla\nabla L$:

$$X_{N+1} \;=\; X_N \;-\; \Big[\, \nabla\nabla L(X_N) + c\,\mathbf{I} \,\Big]^{-1} \nabla L(X_N)\,, \qquad (3.56)$$

where $\mathbf{I}$ is the *identity matrix*, with ones along the diagonal and zeros everywhere else. The reason for this curious suggestion is best understood from the quadratic form in Fig. 3.6, where the nature of a matrix is characterized by its eigenvalues $\{\lambda_j\}$ and eigenvectors $\{e_j\}$; explicitly, for $\nabla\nabla L$, these are the solutions of the equation

$$\big[\nabla\nabla L\big]\, e_j \;=\; \lambda_j\, e_j\,, \qquad (3.57)$$

where $j = 1, 2, \ldots, M$. If we add a multiple of the identity matrix to $\nabla\nabla L$, then we find that this new matrix has the same eigenvectors but different eigenvalues:

$$\big[\nabla\nabla L + c\,\mathbf{I}\big]\, e_j \;=\; \big[\lambda_j + c\big]\, e_j\,, \qquad (3.58)$$

Thus the 'beefing-up of the diagonal' has no effect on the orientation of the ellipsoid and leaves the correlations between the parameters in place; with a suitable choice of c, however, it does cause a significant narrowing in those principal directions in which it was very elongated. Since these small eigenvalues are associated with large uncertainties, eqn (3.56) stabilizes the iterative algorithm by selectively reducing their influence. To put it another way, the inverse of a matrix is related to the reciprocal of its determinant; as this is given by the product of the eigenvalues, the small ones will cause $(\nabla\nabla L)^{-1}$ to blow up. By adding a small (negative) multiple of the identity matrix to $\nabla\nabla L$, we can ensure that the magnitude of the determinant is safely greater than zero; this results in a reduced (finite) value for the inverse of the hybrid second-derivative matrix and gives a correspondingly smaller change in our estimate of the solution between iterations.

If the number of parameters becomes too large, so that the $\nabla\nabla L$ matrix is difficult to store or invert, then the Newton–Raphson procedure can be implemented through a *conjugate-gradient* algorithm; details may be found in *Numerical Recipes* (Press *et al.* 1986), for example. The more serious problem is that the solution to eqn (3.49) will not necessarily represent the position of the maximum, even when the posterior pdf is unimodal. The requirement $\nabla L(X_O) = 0$ is really a condition for finding a stationary point and can sometimes be satisfied in the extreme tails of the posterior probability. The Newton–Raphson procedure will diverge (to infinity), therefore, if the initial guess X_1 is not close enough to the optimal solution. The situation is illustrated schematically in Fig. 3.10.

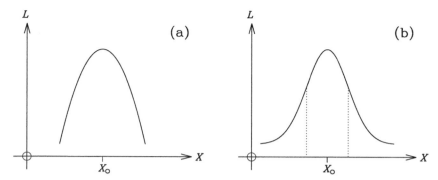

Fig. 3.10 An illustration of the logarithm of two posterior pdfs: (a) a well-behaved (linear-like) problem for which the Newton–Raphson procedure will converge to X_O from any starting point; (b) a unimodal pdf where the Newton–Raphson algorithm will diverge unless the initial guess (X_1) lies within the two dotted lines.

For the case of Fig. 3.10(b), one way to ensure that our computer program will converge towards X_O is to rely predominantly on the direction of the gradient ∇L. A good alternative to such first-derivative methods is provided by the approach of Nelder and Mead (1965): their 'up-hill' *simplex* search algorithm works with the function L directly and is very robust for unimodal pdfs; it is practical to use for up to a couple of dozen parameters. Being gradient-free, it also allows (any) sharp cut-offs imposed by the prior to be encoded in a straightforward manner. Nevertheless, since it lacks the efficiency of Newton–Raphson, it's best used just as a first step to get close to the optimal solution. Besides, we need to evaluate the second-derivative matrix to estimate the error-bars for the parameters; if necessary, the required gradients can even be calculated by *finite-differences* ($\partial L/\partial X_j \approx \left[L(\boldsymbol{X}_{\neq j}, X_j + \delta/2) - L(\boldsymbol{X}_{\neq j}, X_j - \delta/2) \right]/\delta$, etc.).

3.4.4 Hard problems

The most difficult optimization task occurs for the case when the posterior pdf is multimodal. For up to a couple of parameters, or so, the best thing to do is to use brute force and ignorance: simply 'display' the entire pdf, by evaluating it on a discrete grid, and look for the maximum(s). Unfortunately, as mentioned earlier, this becomes hopelessly impractical for more than a handful of variables. While gradient-based algorithms are the most efficient for multivariate analysis, they cannot help for the multimodal case: a knowledge of the local slope sheds no light on the position of the global maximum. To date, all procedures designed to address this problem entail a large element of inspired trial-and-error. *Simulated annealing* is perhaps the most well-established of these techniques and was proposed by Kirkpatrick *et al.* (1983) on the basis of a *thermodynamic* analogy. The basic scheme involves an exploration of the parameter space by a series of trial random changes in the current estimate of the solution; in terms of eqn (3.55), this could be written as $X_{N+1} = X_N + \Delta X$, where ΔX is chosen by a random number generator. The proposed update of the solution is always considered advantageous if it yields a higher value of L, but bad moves are also sometimes accepted. This occasional

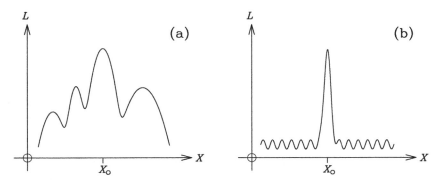

Fig. 3.11 A schematic illustration of the logarithm of two multimodal posterior pdfs. Simulated annealing should work well for case (a), but will not be very helpful for (b).

allowance of retrograde steps provides a mechanism for escaping entrapment in local maxima. The shape of L for two multimodal posterior pdfs is shown in Fig. 3.11. Simulated annealing should work like a treat for case (a), where there is an underlying trend towards a global maximum, but is unlikely to be much more efficient than brute force and ignorance for (b).

The real difficulty with the multimodal case is that it is almost impossible to guarantee that the optimal solution has been found, in a finite amount of time, in general. As such, the field remains wide open. Ingenious new schemes, such as *genetic algorithms* (Michalewicz 1992), are always being put forward, but none can claim universal success or applicability. We must also remember that it doesn't really make sense to talk about a single 'best' answer, even in principle, if the posterior pdf has several maxima of comparable magnitude. In these cases, Monte Carlo methods that can simulate exploration of all the important regions of parameter space will have considerable merit; they are the focus of a more detailed discussion in Chapters 9 and 10.

3.5 Approximations: maximum likelihood and least-squares

There has been one common theme to all our discussions, and examples, so far: namely, the use of Bayes' theorem. Its great virtue lies in the fact that it helps us to relate the pdf required to describe our inference, to others which we have a better chance of being able to assign. Using the components of the vectors X and D to denote the set of M parameters of interest and the N measured data, respectively, we can write Bayes' theorem compactly as

$$\mathrm{prob}(X\,|\,D,I) \propto \mathrm{prob}(D\,|\,X,I) \times \mathrm{prob}(X\,|\,I) \,, \qquad (3.59)$$

where I represents all the relevant background information. The prior pdf should reflect everything we know about X before the analysis of the current data; if we are largely ignorant, then we might indicate this naïvely with a uniform, or flat, pdf (as a simple limiting case of a very broad one):

$$\mathrm{prob}(X\,|\,I) = \mathrm{constant} \,, \qquad (3.60)$$

for, effectively, all values of X. Since this uniform assignment can be absorbed into the (omitted) normalization constant of eqn (3.59), the posterior pdf becomes directly proportional to the likelihood function:

$$\text{prob}(X|D,I) \propto \text{prob}(D|X,I) . \tag{3.61}$$

Thus our best estimate X_0, given by the maximum of the posterior, is equivalent to the solution which yields the greatest value for the probability of the observed data; in this context, it is usually referred to as the *maximum likelihood* estimate.

We can obtain further simplification by making suitable approximations with regard to the likelihood function itself. If we assume that the data are independent, for example, then their joint pdf $\text{prob}(D|X,I)$ is given by the product of the probabilities for the individual measurements:

$$\text{prob}(D|X,I) = \prod_{k=1}^{N} \text{prob}(D_k|X,I) , \tag{3.62}$$

where D_k is the kth datum. Although we have used this result several times before, it is worth emphasizing that it follows immediately from (the repeated application of) the product rule of eqn (1.2),

$$\text{prob}(D_k, D_l|X,I) = \text{prob}(D_k|D_l,X,I) \times \text{prob}(D_l|X,I) ,$$

and the assertion that our knowledge of one datum has no influence on our ability to predict the outcome of another, if we are already given X:

$$\text{prob}(D_k|D_l,X,I) = \text{prob}(D_k|X,I) .$$

If we also assume that the noise associated with the experimental measurements can reasonably be represented as a Gaussian process, then the probability of an individual datum can be written as:

$$\text{prob}(D_k|X,I) = \frac{1}{\sigma_k \sqrt{2\pi}} \exp\left[-\frac{(F_k - D_k)^2}{2\sigma_k^2} \right], \tag{3.63}$$

where I implicitly includes a knowledge of both the expected size of the error-bars $\{\sigma_k\}$, and an adequate model of the functional relationship f between the parameters X and the ideal (noiseless) data F:

$$F_k = f(X,k) . \tag{3.64}$$

Equations (3.62) and (3.63) then allow us to approximate the likelihood function by

$$\text{prob}(D|X,I) \propto \exp\left(-\frac{\chi^2}{2} \right), \tag{3.65}$$

where χ^2 is the sum of the squares of the *normalized residuals* ($R_k = (F_k - D_k)/\sigma_k$):

$$\chi^2 = \sum_{k=1}^{N} \left(\frac{F_k - D_k}{\sigma_k} \right)^2 . \tag{3.66}$$

With the uniform prior of eqn (3.60), according to eqn (3.61), the logarithm L of the posterior pdf is simply given by

$$L = \log_e \left[\text{prob}(\boldsymbol{X} | \boldsymbol{D}, I) \right] = \text{constant} - \frac{\chi^2}{2} . \tag{3.67}$$

Since the maximum of the posterior will occur when χ^2 is smallest, the corresponding optimal solution $\boldsymbol{X}_O$ is usually called the *least-squares* estimate.

Maximum likelihood and least-squares are amongst the most frequently used procedures in data analysis; as we have just seen, they can easily be justified through the Bayesian approach. In this context, however, neither of these methods is seen as fundamental or sacrosanct: they are merely what Bayes' theorem reduces to when certain simplifying approximations are deemed to be suitable. If the results from a least-squares analysis seem unsatisfactory, for example, we don't need to throw-up our hands in horror and say 'Least-squares doesn't work!'. Since we understand the assumptions which underlie its use, it's much more fruitful to ponder whether these were appropriate for the problem being considered; if not, we should start again from Bayes' theorem and derive a better statistical prescription for that case. Some examples of the latter are given in Chapter 8.

Perhaps one of the main reasons for the popularity of the least-squares method is that it is simple to apply. This is particularly true if the functional relationship of eqn (3.64) is linear, because then ∇L can be written in the convenient form of eqn (3.50); as mentioned in Section 3.4, this makes the optimization problem very easy to handle. To see that this is so, let us write the equation for the kth ideal datum as

$$F_k = \sum_{j=1}^{M} T_{kj} X_j + C_k , \tag{3.68}$$

where the values of both T_{kj} and C_k are independent of the parameters $\{X_j\}$; in matrix-vector notation, this would just be $\boldsymbol{F} = \mathsf{T}\boldsymbol{X} + \boldsymbol{C}$. Using the *chain-rule* of differentiation, the jth component of ∇L is given by

$$\frac{\partial L}{\partial X_j} = -\frac{1}{2} \frac{\partial \chi^2}{\partial X_j} = -\sum_{k=1}^{N} \frac{(F_k - D_k)}{\sigma_k^2} \frac{\partial F_k}{\partial X_j} . \tag{3.69}$$

Although we could substitute for F_k, and $\partial F_k / \partial X_j = T_{kj}$, from eqn (3.68) in eqn (3.69), and rearrange it into the form of eqn (3.50), it's slightly messy to do this algebraically. An easier way to verify the linearity of ∇L is to differentiate eqn (3.69) with respect to X_i and notice that the elements of the second-derivative matrix are all constant:

$$\frac{\partial^2 L}{\partial X_i \partial X_j} = -\sum_{k=1}^{N} \frac{T_{ki} T_{kj}}{\sigma_k^2} . \tag{3.70}$$

Since all the higher derivatives of L are identically zero, the posterior pdf is defined completely by the optimal solution X_O and its covariance matrix; the components of the latter are related to twice the inverse of the $\nabla\nabla\chi^2$, or the *Hessian*, matrix:

$$\left\langle (X_i - X_{Oi})(X_j - X_{Oj}) \right\rangle = -\left[(\nabla\nabla L)^{-1} \right]_{ij} = 2\left[(\nabla\nabla\chi^2)^{-1} \right]_{ij}. \qquad (3.71)$$

The useful property that the $\nabla\nabla L$ matrix will be constant if the parameters X are related linearly to the data is not generally true without eqn (3.67); that is why the least-squares approximation is very convenient. For example, the functional model of eqn (3.1), where we are given the shape and position of the signal peak, is linear with respect to the amplitude A and the background B. Nevertheless, the optimal solution (A_O, B_O) is difficult to write down analytically because the gradient-vector of the posterior pdf in eqn (3.8) cannot be rearranged into the form of eqn (3.50); this was not a serious impediment, of course, because we just did the problem numerically (with brute force and ignorance). Even for that case of the Poisson likelihood, however, we could have obtained a fairly good estimate by using the least-squares approximation. This is because eqn (3.2) starts to take on Gaussian characteristics if the expected number of counts is more than a handful ($\geqslant 10$); this behaviour can be seen in Fig. 3.2(b), despite the fact that it's a function of a discrete variable. In the limit of large numbers, it can formally be shown that eqn (3.2) is well-represented by a normal distribution:

$$\text{prob}(N|D) = \frac{D^N e^{-D}}{N!} \propto \exp\left[-\frac{(N-D)^2}{2D} \right],$$

In our usual shorthand notation, we could summarize this by saying $N \approx D \pm \sqrt{D}$. As the number of measured counts will be roughly equal to the expected value D, we can make the denominator of the exponent independent of the parameters A and B by replacing the $\sqrt{D}$ error-bar with $\sqrt{N}$; this helps to linearize the optimization problem. With a uniform prior, the logarithm of the posterior pdf is then approximated well by eqn (3.67); in terms of our present notation, the corresponding χ^2 statistic is given by

$$\chi^2 = \sum_{k=1}^{N} \frac{(F_k - D_k)^2}{D_k},$$

where D_k is now the number of counts measured in the kth data-channel and F_k is our estimate of their expected number based on the linear relationship of eqn (3.1). This result, where the error-bar in eqn (3.66) is replaced by the square root of the datum ($\sigma_k^2 = D_k$, or more often F_k), tallies with the definition of χ^2 found in many elementary textbooks; it suggests that the assumption of a Poisson process is implicit in its use. This would not be appropriate, for example, if the uncertainties related to an experiment were dominated by thermal (Johnson) noise in a piece of electronic equipment; in that case, $\sigma_k =$ a constant (proportional to the temperature) would be more suitable.

Despite the practical benefits of least-squares, we should emphasize that the real justification for its use hinges on the assignments of a Gaussian likelihood function and uniform prior. Otherwise, what reason do we have for using the sum of the squares of the residuals as a misfit statistic as opposed to anything else? Indeed, there is something to be said for the robustness of the l_1-*norm*:

$$l_1\text{-norm} = \sum_{k=1}^{N} \left| \frac{F_k - D_k}{\sigma_k} \right| . \tag{3.72}$$

This sum of the *moduli* of the (normalized) residuals has the advantage that it is far less susceptible to the effects of 'freak data' than is χ^2 (which is called the l_2-*norm*, in this terminology). Although the least-squares prescription was obtained by adopting the traditional assumptions of independent, additive, Gaussian noise in this section, we will see an alternative derivation in Chapter 5; there we will use the maximum entropy principle to assign the relevant uniform, Gaussian and exponential pdfs needed to justify the maximum likelihood, least-squares and l_1-norm estimates.

Before giving a simple example of the use of least-squares, we should warn against the unfortunate connotations conjured up by the words 'maximum likelihood estimate'. In terms of everyday parlance, it suggests that we have obtained the most probable values for the parameters of interest; without qualification, this is not so! By maximizing the likelihood function, we have found the values which make the measured data most probable; although we expect this to be relevant to our real question, 'What are the most probable values of the parameters, given the data?', it is not the same thing. To put it more plainly: $\text{prob}(A|B) \neq \text{prob}(B|A)$, in general. For example, the probability that it will rain given that there are clouds overhead is quite different from the probability of there being clouds overhead given that it's raining. The required relationship between these distinct conceptual entities is, of course, provided by Bayes' theorem. At this juncture, the reader may be forgiven for thinking that this is purely a philosophical point; after all, we have already noted, in eqn (3.61), that the posterior pdf becomes directly proportional to the likelihood function if we assign a uniform prior (which we often do). This belabouring is important, however, because a recognition of the qualification can have practical consequences. Even with a flat prior, a due consideration of the range over which it is valid can lead to a better estimate of the parameters involved and enable us to tackle problems which would otherwise be inaccessible; we will see examples of this in both Section 3.6 and in Chapter 4.

3.5.1 Fitting a straight line

One of the most frequently encountered problems in data analysis is the fitting of a straight line to graphical data. Suppose we are given a set of N data $\{Y_k\}$, with associated error-bars $\{\sigma_k\}$, measured at 'positions' $\{x_k\}$. What is the best estimate of the two parameters of a straight line which describes them?

Let us do this problem as a simple exercise in the use of least-squares. The situation is illustrated schematically in Fig. 3.12. For the straight-line model, the kth ideal datum y_k is given by

$$y_k = m x_k + c ,$$

where m is the slope and c is the intercept. Substituting $F_k = y_k$, and $D_k = Y_k$, into eqn (3.66), we obtain

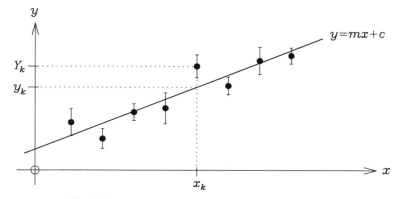

Fig. 3.12 Fitting a straight line to noisy graphical data.

$$\chi^2 = \sum_{k=1}^{N} \frac{(m x_k + c - Y_k)^2}{\sigma_k^2} .$$

According to eqn (3.69), the components of ∇L are given by ($-1/2$ times) the partial derivatives of χ^2:

$$\frac{\partial \chi^2}{\partial m} = \sum_{k=1}^{N} \frac{2 (m x_k + c - Y_k) x_k}{\sigma_k^2} \quad \text{and} \quad \frac{\partial \chi^2}{\partial c} = \sum_{k=1}^{N} \frac{2 (m x_k + c - Y_k)}{\sigma_k^2} .$$

To make the subsequent algebraic manipulations easier, let us rewrite $\nabla\nabla\chi^2$ in the form of eqn (3.50):

$$\nabla\chi^2 = \begin{pmatrix} \alpha & \gamma \\ \gamma & \beta \end{pmatrix} \begin{pmatrix} m \\ c \end{pmatrix} - \begin{pmatrix} p \\ q \end{pmatrix}, \quad (3.73)$$

where the constants α, β, γ, p and q are related to the data through

$$\alpha = \sum w_k x_k^2 , \quad \beta = \sum w_k , \quad \gamma = \sum w_k x_k , \quad p = \sum w_k x_k Y_k , \quad q = \sum w_k Y_k ,$$

in which $w_k = 2/\sigma_k^2$ and the summations are from $k = 1$ to N. The best estimate (m_0, c_0), given by the minimum of χ^2, is defined by the requirement that $\nabla\chi^2 = 0$; the pair of linear simultaneous equations resulting from equating eqn (3.73) to zero can be solved directly, or by matrix inversion as in eqn (3.51), and yield

$$m_0 = \frac{\beta p - \gamma q}{\alpha \beta - \gamma^2} \quad \text{and} \quad c_0 = \frac{\alpha q - \gamma p}{\alpha \beta - \gamma^2} . \quad (3.74)$$

From eqn (3.71), the corresponding covariance matrix for these parameters is given by twice the inverse of $\nabla\nabla\chi^2$. By explicit differentiation, or comparison with eqns (3.50) and (3.52), the latter is the same as the 2×2 matrix in eqn (3.73); thus

$$\begin{pmatrix} \sigma_{mm}^2 & \sigma_{mc}^2 \\ \sigma_{mc}^2 & \sigma_{cc}^2 \end{pmatrix} = 2 \begin{pmatrix} \alpha & \gamma \\ \gamma & \beta \end{pmatrix}^{-1} = \frac{2}{\alpha\beta - \gamma^2} \begin{pmatrix} \beta & -\gamma \\ -\gamma & \alpha \end{pmatrix}, \qquad (3.75)$$

where the square root of the diagonal elements gives the (marginal) error-bars for the inferred values of m and c, and the γ-term describes how they are correlated.

If the error-bars for the data are not known, then we might do the analysis by assuming that they are all of the same size: $\sigma_k = \sigma$. By setting $w_k =$ a constant $= 2/\sigma^2$ in the definitions of α, β, γ, p and q above, we would find that the values of m_o and c_o were independent of the magnitude of σ; however, this would not be true of our estimate of the reliability of the optimal solution. As in Section 3.3, we must take the uncertainty about σ into account by marginalizing over it as a nuisance parameter:

$$\text{prob}(m, c | \{Y_k\}, I) = \int_0^\infty \text{prob}(m, c, \sigma | \{Y_k\}, I) \, d\sigma$$

$$\propto \int_0^\infty \text{prob}(\{Y_k\} | m, c, \sigma, I) \times \text{prob}(m, c, \sigma | I) \, d\sigma,$$

where we have used Bayes' theorem to express the joint posterior pdf on the right as a product of the likelihood function and a prior pdf. With the assumptions of independent, additive, Gaussian noise, the likelihood function is still of the form $\exp(-\chi^2/2)$:

$$\text{prob}(\{Y_k\} | m, c, \sigma, I) \propto \sigma^{-N} \exp\left[-\frac{1}{2\sigma^2} \sum_{k=1}^N (m x_k + c - Y_k)^2 \right],$$

but we now have to be careful to explicitly retain all the factors which depend on σ. If we combine this with the assignment of a uniform prior, then the integral over σ can be evaluated with the same algebra as in Section 3.3; this results in a Student-t distribution, very similar to that of eqn (3.38):

$$\text{prob}(m, c | \{Y_k\}, I) \propto \left[\sum_{k=1}^N (m x_k + c - Y_k)^2 \right]^{-(N-1)/2}.$$

As usual, the best estimate of m and c, and their related covariance matrix, is given by the partial derivatives of the logarithm of this (marginal) posterior pdf. This leads to the recovery of the formulae of eqns (3.74) and (3.75), with the unknown value of the error-bar σ being replaced by an estimate S derived from the data:

$$S^2 = \frac{1}{N-1} \sum_{k=1}^N (m_o x_k + c_o - Y_k)^2.$$

In all the preceding analysis, we have implicitly assumed (in I) that the positions $\{x_k\}$ are known exactly. The case in which there is also a significant uncertainty with regard to the x-coordinate is more challenging, and we refer the avid reader to Gull (1989*b*) for a discussion of the subtleties involved.

3.6   Error-propagation: changing variables

The last topic in this chapter is often called the propagation of errors; despite the somewhat alarming title, suggestive of an unfortunate amplification of mistakes, it's quite
innocuous. Here we are concerned with the question of how uncertainties in our estimate of a set of parameters translate into reliabilities of quantities derived from them.
For example, suppose we are told that $X = 10 \pm 3$ and $Y = 7 \pm 2$; what can we say about
the difference $X - Y$, or the ratio X/Y, or the sum of their squares $X^2 + Y^2$, and so
on? In essence, the problem is nothing more than an exercise in the change of variables:
given the pdf $\mathrm{prob}(X, Y | I)$, where the relevant information I includes a knowledge
of the data if dealing with the posterior, we need the corresponding pdf $\mathrm{prob}(Z | I)$,
where $Z = X - Y$, or $Z = X/Y$, or whatever, as appropriate.

Let us start with the easiest type of transformation; namely, one that involves a single
variable and some function of it. Given that $Y = f(X)$, how is the pdf $\mathrm{prob}(X | I)$
related to $\mathrm{prob}(Y | I)$? Imagine taking a very small interval δX about some arbitrary
point $X = X^*$; the probability that X lies in the range $X^* - \delta X/2$ to $X^* + \delta X/2$ is
given by

$$\mathrm{prob}\left(X^* - \frac{\delta X}{2} \leqslant X < X^* + \frac{\delta X}{2} \,\middle|\, I\right) \approx \mathrm{prob}(X = X^* | I)\,\delta X\,, \qquad (3.76)$$

where the equality becomes exact in the limit $\delta X \to 0$. Now suppose that we view this
pdf as a function of another quantity Y, related (monotonically) to X by $Y = f(X)$;
then, f will map the point $X = X^*$ (uniquely) to $Y = Y^* = f(X^*)$ and the interval
δX to the corresponding region δY. The situation is shown schematically in Fig. 3.13.
Since the range of Y values spanned by $Y^* \pm \delta Y/2$ is equivalent to a variation in X
between $X^* \pm \delta X/2$, the area under the pdf $\mathrm{prob}(Y | I)$ should equal the probability
represented by eqn (3.76). Therefore, we require that

$$\mathrm{prob}(X = X^* | I)\,\delta X \;=\; \mathrm{prob}(Y = Y^* | I)\,\delta Y\,.$$

As this must be true for any point in X-space then, in the limit of infinitesimally small
intervals, we obtain the relationship

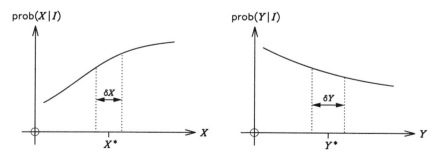

Fig. 3.13 Changing variables in one dimension: the function f maps the point X^* to $Y^* = f(X^*)$
and the small interval δX to the corresponding region δY.

$$\text{prob}(X|I) \;=\; \text{prob}(Y|I) \times \left| \frac{\mathrm{d}Y}{\mathrm{d}X} \right|, \tag{3.77}$$

where the term on the far right, given by the modulus of $\mathrm{d}f/\mathrm{d}X$, is called the *Jacobian*. The reason for taking the magnitude of the derivative is to ensure that it represents a ratio of lengths even when positive increments in X map to negative ones for Y. If $Y = f(X)$ is not a one-to-one transformation, then eqn (3.77) has to be extended to a summation over all the Y values which correspond to a given X.

As a concrete example of this procedure, let's return to the lighthouse problem of Section 2.4. There, in eqn (2.32), we had assigned a uniform pdf for the azimuth θ: $\text{prob}(\theta|\alpha, \beta, I) = 1/\pi$, where the angle θ had to lie between $\pm\pi/2$ radians. In eqn (2.33), we also had the connection between the azimuth and the position along the coast x: $\beta \tan\theta = x - \alpha$. Differentiating both sides of eqn (2.33) with respect to x, we have

$$\beta \sec^2\theta \times \frac{\mathrm{d}\theta}{\mathrm{d}x} = 1.$$

Using the trigonometric identity $\tan^2\theta + 1 \equiv \sec^2\theta$, and substituting for $\tan\theta$ from eqn (2.33), the Jacobian can be written as

$$\frac{\mathrm{d}\theta}{\mathrm{d}x} \;=\; \left(\beta\left[1 + \tan^2\theta\right] \right)^{-1} \;=\; \left(\beta\left[1 + \left(\frac{x-\alpha}{\beta}\right)^2\right] \right)^{-1}.$$

Finally, we can use eqn (3.77) to transform the pdf for θ in eqn (2.32) to its equivalent form in terms of x; after a little algebraic rearrangement, we obtain (as promised) the Cauchy distribution of eqn (2.34):

$$\text{prob}(x|\alpha, \beta, I) \;=\; \text{prob}(\theta|\alpha, \beta, I) \times \left| \frac{\mathrm{d}\theta}{\mathrm{d}x} \right| \;=\; \frac{\beta}{\pi\left[\beta^2 + (x-\alpha)^2\right]}.$$

The result in eqn (3.77) can be generalized to the case of several variables. Although it becomes more difficult to give a simple pictorial representation, like the one in Fig. 3.13, the central theme of the argument is the same: if we want to write the pdf for M parameters $\{X_j\}$ in terms of the same number of quantities $\{Y_j\}$ related to them, then we must ensure that

$$\text{prob}(\{X_j\}|I)\, \delta X_1 \delta X_2 \cdots \delta X_M \;=\; \text{prob}(\{Y_j\}|I)\, \delta^M \text{Vol}(\{Y_j\}),$$

where $\delta^M \text{Vol}(\{Y_j\})$ is the M-dimensional volume in Y-space mapped out by the small hypercube region $\delta X_1 \delta X_2 \cdots \delta X_M$ in X-space. After some effort, most textbooks on mathematical methods (for scientists) derive the formula

$$\delta^M \text{Vol}(\{Y_j\}) \;=\; \left| \frac{\partial\,(Y_1, Y_2, \cdots, Y_M)}{\partial\,(X_1, X_2, \cdots, X_M)} \right| \delta X_1 \delta X_2 \cdots \delta X_M,$$

where the strange-looking quantity in the modulus sign is the multivariate Jacobian; it is given by the determinant of the $M \times M$ matrix of partial derivatives $\partial Y_i/\partial X_j$. Thus the general form of eqn (3.77) can be written as

$$\text{prob}(\{X_j\}|I) \;=\; \text{prob}(\{Y_j\}|I) \times \left| \frac{\partial (Y_1, Y_2, \cdots, Y_M)}{\partial (X_1, X_2, \cdots, X_M)} \right|. \tag{3.78}$$

To illustrate the use of eqn (3.78), let's consider the transformation of a pdf defined on a two-dimensional *Cartesian* grid (x, y) to its equivalent form in *polar* coordinates (R, θ). From Fig. 3.14(a), it can be seen that the functional relationship between the two sets of parameters is given by

$$x = R\cos\theta \qquad \text{and} \qquad y = R\sin\theta \,.$$

Taking the partial derivatives of x and y with respect to R and θ, we can easily evaluate the determinant of the resulting 2×2 matrix for the Jacobian:

$$\left| \frac{\partial (x, y)}{\partial (R, \theta)} \right| = \left| \begin{matrix} \cos\theta & -R\sin\theta \\ \sin\theta & R\cos\theta \end{matrix} \right| = R\left[\cos^2\theta + \sin^2\theta \right] = R \,,$$

where the final simplification is obtained from the identity $\sin^2\theta + \cos^2\theta \equiv 1$. According to eqn (3.78), therefore, the pdf $\text{prob}(R, \theta|I)$ is related to $\text{prob}(x, y|I)$ by

$$\text{prob}(R, \theta|I) = \text{prob}(x, y|I) \times R \,. \tag{3.79}$$

Thus if the pdf for x and y was an *isotropic*, bivariate, Gaussian,

$$\text{prob}(x, y|I) = \frac{1}{2\pi\sigma^2} \exp\left[-\frac{(x^2 + y^2)}{2\sigma^2} \right], \tag{3.80}$$

then the corresponding pdf for R and θ would take the form

$$\text{prob}(R, \theta|I) = \frac{R}{2\pi\sigma^2} \exp\left(-\frac{R^2}{2\sigma^2} \right). \tag{3.81}$$

Rather than working through the formal Jacobian analysis given above, we could have obtained eqn (3.79) directly from a simple geometrical argument: the probability

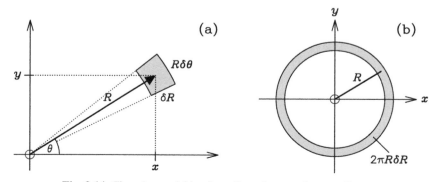

Fig. 3.14 Changing variables from Cartesian to polar coordinates.

that the polar parameters lie in the small range $R \pm \delta R/2$, and $\theta \pm \delta\theta/2$, is given by the product of the value of the pdf $\mathrm{prob}(x, y | I)$ at the point corresponding to (R, θ) and the shaded element of area $R \, \delta\theta \, \delta R$ in Fig. 3.14(a); by the definition of the pdf for R and θ, of course, this is equal to $\mathrm{prob}(R, \theta | I) \, \delta\theta \, \delta R$. Hence, the desired result follows: $\mathrm{prob}(R, \theta | I) \, \delta R \, \delta\theta = \mathrm{prob}(x, y | I) \, R \, \delta R \, \delta\theta$.

Having transformed $\mathrm{prob}(x, y | I)$ to polar coordinates, it becomes a straightforward task to obtain the pdf for the radius $R = \sqrt{x^2 + y^2}$: we just need to marginalize the joint pdf $\mathrm{prob}(R, \theta | I)$ over θ. For the case of the bivariate Gaussian in eqn (3.80), eqn (3.81) is integrated readily with respect to θ:

$$\mathrm{prob}(R | I) = \int_0^{2\pi} \mathrm{prob}(R, \theta | I) \, \mathrm{d}\theta = \frac{R}{\sigma^2} \exp\left(-\frac{R^2}{2\sigma^2}\right). \qquad (3.82)$$

Again, we could also have arrived at this result pictorially: since the value of the pdf in eqn (3.80) depends only on the distance from the origin, and not the direction, the probability that R lies in a narrow range δR is given by the product of the magnitude of the Gaussian pdf at that radius and the corresponding area of the shaded ring shown in Fig. 3.14(b); hence, $\mathrm{prob}(R | I) \, \delta R = \mathrm{prob}(x, y | I) \, 2\pi R \, \delta R$ and eqn (3.82) follows immediately with the substitution of $x^2 + y^2 = R^2$ in eqn (3.80).

A multidimensional generalization of this last argument allows us to derive the χ^2 distribution mentioned towards the end of Section 3.3. The likelihood function of eqns (3.65) and (3.66) is an N-dimensional, isotropic, Gaussian when viewed in terms of the normalized residuals:

$$\mathrm{prob}(\boldsymbol{D} | \boldsymbol{X}, I) \propto \exp\left[-\frac{r_1^2 + r_2^2 + \cdots + r_N^2}{2}\right],$$

where $r_k = (F_k - D_k)/\sigma_k$. As in the case above, the value of the pdf depends only on the distance from the origin ($r_k = 0$); this radius $R = \sqrt{\sum r_k^2}$ is, of course, just the square root of χ^2. The probability that R lies in a narrow range δR, $\mathrm{prob}(R | \boldsymbol{X}, I) \, \delta R$, will equal the product of the magnitude of the likelihood function at that radius and the hypervolume of the associated spherical shell; since the latter is proportional to R^{N-1}, the marginal distribution is just an extension of eqn (3.82):

$$\mathrm{prob}(R | \boldsymbol{X}, I) \propto R^{N-1} \exp\left(-R^2/2\right).$$

To convert this into a pdf for χ^2, we just need to carry out a one-to-one transformation according to eqn (3.77); using the functional relationship $\chi^2 = R^2$, we obtain

$$\mathrm{prob}(\chi^2 | \boldsymbol{X}, I) \propto \left(\chi^2\right)^{N/2 - 1} \exp\left(-\chi^2/2\right). \qquad (3.83)$$

Technically, this is called a χ^2 distribution with N degrees of freedom. For $N \geqslant 2$, it has a maximum at $N - 2$; if the number of data is large, its shape is well-described by a Gaussian pdf summarized by $\chi^2 \approx N \pm \sqrt{2N}$.

We have now seen the basic ingredients required for the propagation of errors: it either involves a transformation in the sense of eqn (3.78), or an integration such as eqn

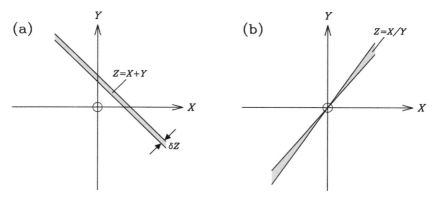

Fig. 3.15 The elements of area over which the joint pdf for the two parameters $\mathrm{prob}(X,Y|I)$ must be integrated to obtain an estimate of: (a) the sum $Z=X+Y$; and (b) the ratio $Z=X/Y$.

(3.82), or a combination of the two. For the common problem of wanting to estimate the sum $Z=X+Y$, or the ratio $Z=X/Y$, of two parameters X and Y, we simply need to integrate the joint pdf $\mathrm{prob}(X,Y|I)$ along the shaded strips in Fig. 3.15. If this is not intuitively obvious, we can justify it analytically by using marginalization and the product rule:

$$\mathrm{prob}(Z|I) \;=\; \iint \mathrm{prob}(Z|X,Y,I)\,\mathrm{prob}(X,Y|I)\,\mathrm{d}X\,\mathrm{d}Y$$

$$=\; \iint \delta\big(Z-f(X,Y)\big)\,\mathrm{prob}(X,Y|I)\,\mathrm{d}X\,\mathrm{d}Y, \qquad (3.84)$$

where the Dirac δ-function in the second line is equal to zero unless $Z=f(X,Y)$, and indicates that Z is determined unambiguously (by the function f) when X and Y are known. Let's explicitly work through this procedure for the case $Z=X+Y$.

Following eqn (3.84), the integral along the diagonal strip in Fig. 3.15(a) can be written as

$$\mathrm{prob}(Z|I) \;=\; \iint \mathrm{prob}(X,Y|I)\,\delta\big(Z-(X+Y)\big)\,\mathrm{d}X\,\mathrm{d}Y. \qquad (3.85)$$

If the information I only tells us that $X = x_\mathrm{o} \pm \sigma_x$, and $Y = y_\mathrm{o} \pm \sigma_y$, then it is reasonable to assume that these parameters are not correlated; $\mathrm{prob}(X,Y|I)$ is then just the product of the individual pdfs for X and Y, so that eqn (3.85) becomes

$$\mathrm{prob}(Z|I) \;=\; \int \mathrm{d}X\,\mathrm{prob}(X|I) \int \mathrm{prob}(Y|I)\,\delta\big(Z-X-Y\big)\,\mathrm{d}Y.$$

Since the δ-function is infinitely sharp (but has unit area), the Y integrand is zero unless $Y=Z-X$; the integral with respect to Y is trivial, therefore, and the pdf for the sum reduces to a *convolution* of the pdfs for X and Y:

$$\text{prob}(Z|I) = \int \text{prob}(X|I)\,\text{prob}(Y = Z - X\,|I)\,\mathrm{d}X\,. \qquad (3.86)$$

The nature of the information I also suggests that we should assign Gaussian pdfs for the two parameters, with maxima at x_0 and y_0, and widths σ_x and σ_y, respectively; substituting these into eqn (3.86), we have

$$\text{prob}(Z|I) = \frac{1}{2\pi\sigma_x\sigma_y} \int_{-\infty}^{+\infty} \exp\left[-\frac{(X - x_0)^2}{2\sigma_x^2}\right] \exp\left[-\frac{(Z - X - y_0)^2}{2\sigma_y^2}\right] \mathrm{d}X\,.$$

After some fairly tedious algebra, consisting largely of completing the square for X in the exponent and simplifying the resultant expression, we obtain

$$\text{prob}(Z|I) = \frac{1}{\sigma_z\sqrt{2\pi}} \exp\left[-\frac{(Z - z_0)^2}{2\sigma_z^2}\right]\,, \qquad (3.87a)$$

where

$$z_0 = x_0 + y_0 \quad \text{and} \quad \sigma_z^2 = \sigma_x^2 + \sigma_y^2\,. \qquad (3.87b)$$

Thus the pdf for the sum is a Gaussian, with a maximum at z_0 and width of σ_z; in fact, the pdf for the difference $Z = X - Y$ is the same except that z_0 is then given by $x_0 - y_0$.

3.6.1 A useful short cut

The calculation above seems rather long-winded for such a simple problem. Could we have got there by a simpler route? After all, we are not usually interested in knowing the fine detail of the shapes of the pdfs; for practical purposes, we're often satisfied to approximate them with Gaussians. Within such limits, the answer is 'Yes, there is an easier method!'

Intuitively, we might have guessed that the best estimate for the difference of the two parameters was $x_0 - y_0$; the corresponding error-bar, given σ_x and σ_y, requires a little more thought. Suppose we perturb $Z = X - Y$:

$$\delta Z = \delta X - \delta Y\,. \qquad (3.88)$$

This would then tell us how small changes in the values of X and Y would affect our estimate of their difference. For the error-bars, of course, we are interested in studying the deviations from the optimal solution: $\delta X = X - x_0$, and so on. According to the integral definitions of eqns (3.25) and (3.26), and the information I, we know that

$$\langle \delta X^2 \rangle = \sigma_x^2\,, \qquad \langle \delta Y^2 \rangle = \sigma_y^2\,, \qquad \langle \delta X\,\delta Y \rangle = 0\,. \qquad (3.89)$$

Note that by δX^2 we mean $(\delta X)^2$, not $\delta(X^2)$, and so on. Squaring both sides of eqn (3.88) and taking expectation values, we have

$$\langle \delta Z^2 \rangle = \langle \delta X^2 + \delta Y^2 - 2\,\delta X\,\delta Y \rangle = \langle \delta X^2 \rangle + \langle \delta Y^2 \rangle - 2\langle \delta X\,\delta Y \rangle\,,$$

where we have used the linear property that the integral of a sum of terms is equal to the sum of the integrals of those terms, in separating the expectations in the last step. Substituting from eqn (3.89), we obtain the error-bar of the difference $X - Y$:

$$\sigma_z = \sqrt{\langle \delta Z^2 \rangle} = \sqrt{\sigma_x^2 + \sigma_y^2} \, .$$

As promised, this is the same as for the sum $X + Y$ in eqn (3.87b).

For a second example of this procedure, let's work out the formula for the error-bar of the ratio of two parameters. Perturbing $Z = X/Y$, using the *quotient* rule of differentiation, we have

$$\delta Z = \frac{Y \, \delta X - X \, \delta Y}{Y^2} \, .$$

This expression can be simplified slightly by dividing through by the ratio itself:

$$\frac{\delta Z}{Z} = \frac{\delta X}{X} - \frac{\delta Y}{Y} \, .$$

Squaring both sides and taking expectation values, we obtain

$$\frac{\langle \delta Z^2 \rangle}{z_0^2} = \frac{\langle \delta X^2 \rangle}{x_0^2} + \frac{\langle \delta Y^2 \rangle}{y_0^2} - 2 \frac{\langle \delta X \, \delta Y \rangle}{x_0 \, y_0} \, ,$$

where the X, Y and Z in the denominator have been replaced by the constants x_0, y_0 and z_0 ($= x_0/y_0$) because we are interested in the deviations from the optimal solution. Finally, substituting from eqn (3.89) and taking square roots gives the desired formula for the error-bar of the ratio:

$$\frac{\sigma_z}{z_0} = \sqrt{\left(\frac{\sigma_x}{x_0} \right)^2 + \left(\frac{\sigma_y}{y_0} \right)^2} \, . \tag{3.90}$$

Equivalently, changing variables to logarithms through eqn (3.77) gives us

$$\sigma_{\log Z} = \sqrt{\sigma_{\log X}^2 + \sigma_{\log Y}^2} \, ,$$

which illustrates the variability of $\log Z = \log X - \log Y$. Despite its virtues, let us end our discussion of error-propagation with a salutary warning against the blind use of this nifty short cut.

3.6.2 Taking the square root of a number

Data of the type shown in Fig. 3.3 are often collected in diffraction studies of crystalline materials. With reference to the associated model of Fig. 3.1, the amplitude of the Bragg (or signal) peak is usually estimated using a least-squares fitting program and the results stated as $A = A_0 \pm \sigma_A$; any uncertainties about the background and the peak shape do, of course, give rise to a correspondingly larger (marginal) error-bar. Since this amplitude is related to the modulus squared of the *complex structure factor*, $A = |F|^2$, crystallographers need to take square roots. Writing $f = |F|$, therefore, we have

$$A = f^2. \tag{3.91}$$

For the purposes of this simple example, we are merely interested in obtaining the best estimate f_o and a measure of its reliability σ_f.

This seems like a fairly easy problem for the short cut procedure: the optimal solution is obviously $f_o = \sqrt{A_O}$; to propagate the error, we just differentiate eqn (3.91), square both sides, and take the expectation values:

$$\langle \delta A^2 \rangle = 4 f_o^2 \langle \delta f^2 \rangle = 4 A_O \langle \delta f^2 \rangle,$$

Substituting $\sigma_A^2 = \langle \delta A^2 \rangle$, and rearranging this equation for the error-bar $\sigma_f = \sqrt{\langle \delta f^2 \rangle}$, we quickly arrive at the result

$$f = \sqrt{A_O} \pm \frac{\sigma_A}{2\sqrt{A_O}}. \tag{3.92}$$

Unfortunately, this elementary analysis breaks down when A_O is negative! Such an occurrence is not unusual, particularly for weak and strongly overlapping *reflections* (or signal peaks). So, what has gone wrong?

Well, we have made two mistakes — both are quite common. The first stems from our failure to distinguish clearly between the posterior pdf and the likelihood function. The least-squares fit is just a shorthand way of saying that, as a function of the amplitude, $\mathrm{prob}(\{data\}|A, I)$ is approximately Gaussian, with a maximum at A_O and a width of σ_A:

$$\mathrm{prob}(\{data\}|A, I) \propto \exp\left[-\frac{(A - A_O)^2}{2\sigma_A^2}\right]. \tag{3.93}$$

Our inference about the value of A, however, is described by $\mathrm{prob}(A|\{data\}, I)$. To relate these two pdfs, we need to use Bayes' theorem:

$$\mathrm{prob}(A|\{data\}, I) \propto \mathrm{prob}(\{data\}|A, I) \times \mathrm{prob}(A|I). \tag{3.94}$$

Since we know physically that the amplitude must be positive, we should, at least, assign a prior which is zero for negative values of A; a naïve choice which incorporates this information is

$$\mathrm{prob}(A|I) = \begin{cases} \text{constant} & \text{for } A \geqslant 0, \\ 0 & \text{otherwise}. \end{cases} \tag{3.95}$$

Multiplying eqns (3.93) and (3.95), we see that the best estimate of A will always be positive. This is true even when the amplitude A_O which gives the closest agreement with the data is negative; in that case, the posterior pdf will be a severely truncated Gaussian. The short cut procedure for the error-propagation will be flawed, therefore, because it implicitly relies on an expansion about a centrally located maximum; this was our second mistake. Since there are no such limitations on the formal change of variables, from $\mathrm{prob}(A|\{data\}, I)$ to $\mathrm{prob}(f|\{data\}, I)$, we should do the proper calculation according to eqn (3.77).

Having obtained the posterior pdf for the amplitude of the Bragg peak, the transformation we require to describe our inference about the modulus of the structure factor is given by

$$\mathrm{prob}(f|\{data\}, I) \;=\; \mathrm{prob}(A|\{data\}, I) \times \left| \frac{\mathrm{d}A}{\mathrm{d}f} \right|,$$

where the derivative of eqn (3.91) yields the Jacobian $|\mathrm{d}A/\mathrm{d}f| = 2f$, with $f = |F| \geqslant 0$. This allows us to write the posterior pdf for f as

$$\mathrm{prob}(f|\{data\}, I) \;\propto\; f \times \exp\left[-\frac{\left(f^2 - A_0\right)^2}{2\,\sigma_A^2} \right], \tag{3.96}$$

for $f \geqslant 0$, and zero otherwise. As usual, the strange-looking expression of eqn (3.96) can be approximated by the more familiar Gaussian pdf:

$$\mathrm{prob}(f|\{data\}, I) \;\approx\; \frac{1}{\sigma_f \sqrt{2\pi}} \; \exp\left[-\frac{\left(f - f_0\right)^2}{2\,\sigma_f^2} \right], \tag{3.97}$$

where, according to eqns (2.12) and (2.15), the parameters f_0 and σ_f are given by the first and second derivatives of the logarithm $L = \log_e\left[\mathrm{prob}(f|\{data\}, I)\right]$:

$$2\,f_0^2 \;=\; A_0 + \left(A_0^2 + 2\sigma_A^2\right)^{1/2}, \tag{3.98a}$$

where

$$\sigma_f^{-2} \;=\; \frac{1}{f_0^2} + \frac{2\left(3 f_0^2 - A_0\right)}{\sigma_A^2}. \tag{3.98b}$$

It can easily be shown that these formulae for the best estimate of f, and its error-bar, reduce to the form of eqn (3.92) in the limit $A_0 \gg \sigma_A$. For those reflections where the amplitude of the Bragg peak is poorly determined by the data, the answers can be very different. Let us illustrate this graphically with a few examples.

Consider first of all a good reflection, $A = 9 \pm 1$, as depicted in the top panels in Fig. 3.16. It confirms that, in this case, the result of the short cut procedure agrees with the proper solution. Moving on to the next pair of pdfs, which are for a poorly determined amplitude, $A = 1 \pm 9$, we notice an interesting divergence. The solution of eqn (3.96), which is plotted with a solid line, indicates that the magnitude of the structure factor could be vanishingly small but is very unlikely to be larger than 6; the optimal estimate is about 2.6. By contrast, the result of eqn (3.92), shown as a dashed line, gives 1.0 for the best estimate and does not exclude values of up to about 12 as being unreasonable. It can also be seen that the Gaussian pdf of eqns (3.97) and (3.98), plotted with a dotted line, is a respectable approximation to eqn (3.96). Finally, the most striking case occurs when A_0 is negative; for example, $A = -20 \pm 9$. The last two panels in Fig. 3.16 show that the proper probabilistic solution is perfectly well-behaved, whereas the short cut

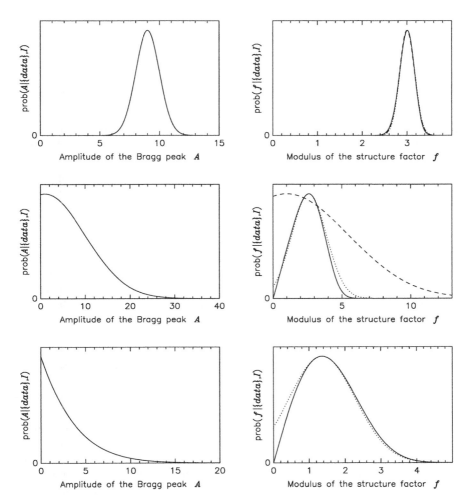

Fig. 3.16 The left-hand panels show the posterior pdf for the amplitude of the Bragg peak, given the naïve prior of eqn (3.95) and the least-squares likelihood function of eqn (3.93), with $A = 9 \pm 1$, 1 ± 9 and -20 ± 9. The right-hand plots are the corresponding pdfs for the modulus of the structure factor ($f = \sqrt{A}$): the solid, dotted and (when appropriate) dashed lines are for eqns (3.96), (3.98) and (3.92), respectively.

procedure breaks down completely because eqn (3.92) does not yield a real value for f. Thus we see how a simple recognition of the relevance of the prior, and strict adherence to the rules of probability theory, can offer a significant advantage over a 'cook-book' approach to least-squares and error-propagation. This analysis can be extended to the case of strongly overlapping peaks, which often occurs in powder diffraction; details can be found in Sivia and David (1994).

4 Model selection

So far, we have been concerned with the problem of parameter estimation. In studying the linear relationship between two quantities, for example, we discussed how to infer the gradient and intercept of the associated straight line. Often, however, there is a question as to whether a quadratic, or even cubic, function might be more appropriate. In this chapter, we go on to consider such cases when there is uncertainty as to which one of a set of alternative models is most suitable.

4.1 Introduction: the story of Mr A and Mr B

In the analysis of the data in Fig. 3.3, we took the signal peak to be Gaussian in shape and the background as flat; this was reasonable since the experimental counts were generated in a computer simulation designed to conform to this description. In real life, of course, we would have to choose a functional form based on the relevant background information available. This could include theoretical considerations, the results of calibration measurements or merely an approximation to simplify the algebra; in all cases, the underlying assumptions need to be stated clearly (in the conditioning on I). Suppose, however, there was real debate as to whether the signal peak should be Gaussian or Lorentzian. How can we decide which is better?

The type of question posed above is often called model selection, or model comparison; we will see several different examples of it in this chapter. Naïvely, we might think that a choice between proposed alternatives can be made on the basis of how well they fit the data. A little reflection soon reveals a potential difficulty in that more complicated models, defined by many parameters, will always be able to give better agreement with the experimental measurements. Although a tenth-order polynomial might yield a closer match with one-dimensional graphical data than a simple straight line, most people would prefer the latter unless the discrepancy was very large. Since an analysis of such problems soon becomes cluttered with algebraic detail, to take into account the varying degrees of flexibility allowed by the different models, let us begin with an elementary formulation due to Jeffreys (1939); following Gull's presentation (1988), we call it the story of Mr A and Mr B.

Mr A has a theory; Mr B also has a theory, but with an adjustable parameter λ.
Whose theory should we prefer on the basis of data D?

Despite its humorous overtones, this represents the bare bones of the model selection question. In the context of our graph-fitting example, Mr A could be the person who thinks that the (noisy) measurements of y against x are described by $y = 0$; Mr B believes that they pertain to $y = a$, but is not sure about the value of the constant a; there could also be a Mr C who is willing to allow the possibility of a non-zero slope, so that $y = a + bx$, and therefore has two adjustable coefficients; and so on. The easiest comparison is that between Mr A and Mr B, since it involves only one unknown parameter; the extension to other problems entails a multivariate generalization of the arguments below, but the basic principles are the same.

From our discussion in Chapter 1, it is clear that we need to evaluate the posterior probabilities for A and B being correct to ascertain the relative merit of the two theories. If the ratio of the posteriors,

$$\text{posterior ratio} \;=\; \frac{\text{prob}(\text{A}\,|\,D,I)}{\text{prob}(\text{B}\,|\,D,I)} \,, \tag{4.1}$$

is very much greater than one, then we will prefer A's theory; if it is very much less than one, then we prefer that of B; and if it is of order unity, then the current data are insufficient to make an informed judgement.

To estimate the odds in eqn (4.1), let us start by applying Bayes' theorem to both the numerator and the denominator; this gives

$$\frac{\text{prob}(\text{A}\,|\,D,I)}{\text{prob}(\text{B}\,|\,D,I)} \;=\; \frac{\text{prob}(D\,|\,\text{A},I)}{\text{prob}(D\,|\,\text{B},I)} \times \frac{\text{prob}(\text{A}\,|\,I)}{\text{prob}(\text{B}\,|\,I)} \,, \tag{4.2}$$

because the term $\text{prob}(D\,|\,I)$ cancels out, top and bottom. As usual, probability theory warns us immediately that the answer to our question depends partly on what we thought about the two theories before the analysis of the data. To be fair, we might take the ratio of the prior terms, on the far right of eqn (4.2), to be unity; a harsher assignment could be based on the past track records of the theorists! To assign the probabilities involving the experimental measurements, $\text{prob}(D\,|\,\text{A},I)$ and $\text{prob}(D\,|\,\text{B},I)$, we need to be able to compare the data with the predictions of A and B: the larger the mismatch, the lower the corresponding probability. This calculation is straightforward for Mr A, but not for Mr B; the latter cannot make predictions without a value for λ.

To circumvent this difficulty, we can use the sum and product rule to relate the probability we require to other pdfs which might be easier to assign. In particular, marginalization and the product rule allow us to express $\text{prob}(D\,|\,\text{B},I)$ as

$$\text{prob}(D\,|\,\text{B},I) \;=\; \int \text{prob}(D,\lambda\,|\,\text{B},I)\,\mathrm{d}\lambda$$

$$=\; \int \text{prob}(D\,|\,\lambda,\text{B},I)\,\text{prob}(\lambda\,|\,\text{B},I)\,\mathrm{d}\lambda\,. \tag{4.3}$$

The first term in the integral $\text{prob}(D\,|\,\lambda,\text{B},I)$, where the value of λ is given, is now just an ordinary likelihood function; as such, it is on a par with $\text{prob}(D\,|\,\text{A},I)$. The second term is B's prior pdf for λ; the onus is, therefore, on that theorist to articulate his state of knowledge, or ignorance, before he is given access to the data.

To proceed further analytically, let us make some simplifying approximations. Assume that, *a priori*, Mr B is only prepared to say that λ must lie between the limits $\lambda_{\min}$ and $\lambda_{\max}$; we can then naïvely assign a uniform prior within this range:

$$\mathrm{prob}(\lambda|\mathrm{B},I) \;=\; \frac{1}{\lambda_{\mathrm{max}} - \lambda_{\mathrm{min}}} \qquad \text{for } \lambda_{\mathrm{min}} \leqslant \lambda \leqslant \lambda_{\mathrm{max}}, \qquad (4.4)$$

and zero otherwise. Let us also take it that there is a value λ_{o} which yields the closest agreement with the measurements; the corresponding probability $\mathrm{prob}(D|\lambda_{\mathrm{o}},\mathrm{B},I)$ will be the maximum of B's likelihood function. As long as this adjustable parameter lies in the neighbourhood of the optimal value, $\lambda_{\mathrm{o}} \pm \delta\lambda$, we would expect a reasonable fit to the data; this can be represented by the Gaussian pdf

$$\mathrm{prob}(D|\lambda,\mathrm{B},I) \;=\; \mathrm{prob}(D|\lambda_{\mathrm{o}},\mathrm{B},I) \times \exp\!\left[-\frac{(\lambda - \lambda_{\mathrm{o}})^2}{2\,\delta\lambda^2} \right]. \qquad (4.5)$$

The assignments of eqns (4.4) and (4.5) are illustrated in Fig. 4.1. We may note that, unlike the prior pdf $\mathrm{prob}(\lambda|\mathrm{B},I)$, B's likelihood function need not be normalized with respect to λ; in other words, $\mathrm{prob}(D|\lambda_{\mathrm{o}},\mathrm{B},I)$ need not equal $1/(\delta\lambda\sqrt{2\pi})$. This is because the λ in $\mathrm{prob}(D|\lambda,\mathrm{B},I)$ appears in the conditioning statement, whereas the normalization requirement applies to quantities to the left of the ' | ' symbol.

In the evaluation of $\mathrm{prob}(D|\mathrm{B},I)$, we can make use of the fact that the prior of eqn (4.4) does not depend explicitly on λ; this enables us to take $\mathrm{prob}(\lambda|\mathrm{B},I)$ outside the integral in eqn (4.3):

$$\mathrm{prob}(D|\mathrm{B},I) \;=\; \frac{1}{\lambda_{\mathrm{max}} - \lambda_{\mathrm{min}}} \int_{\lambda_{\mathrm{min}}}^{\lambda_{\mathrm{max}}} \mathrm{prob}(D|\lambda,\mathrm{B},I)\,\mathrm{d}\lambda, \qquad (4.6)$$

having set the limits according to the specified range. Assuming that the sharp cut-offs at λ_{min} and λ_{max} do not cause a significant truncation of the Gaussian pdf in eqn (4.5), its integral will be equal to $\delta\lambda\sqrt{2\pi}$ times $\mathrm{prob}(D|\lambda_{\mathrm{o}},\mathrm{B},I)$. The troublesome term then reduces to

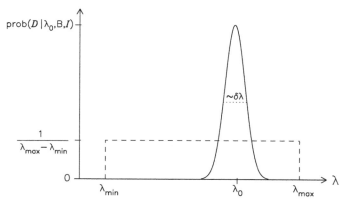

Fig. 4.1 A schematic representation of the prior pdf (dashed line) and the likelihood function (solid line) for the parameter λ in Mr B's theory.

$$\text{prob}(D|\text{B},I) = \frac{\text{prob}(D|\lambda_\text{o},\text{B},I) \times \delta\lambda \sqrt{2\pi}}{\lambda_\text{max} - \lambda_\text{min}}. \quad (4.7)$$

Substituting this into eqn (4.2), we finally see that the ratio of the posteriors required to answer our original question decomposes into the product of three terms:

$$\frac{\text{prob}(\text{A}|D,I)}{\text{prob}(\text{B}|D,I)} = \frac{\text{prob}(\text{A}|I)}{\text{prob}(\text{B}|I)} \times \frac{\text{prob}(D|\text{A},I)}{\text{prob}(D|\lambda_\text{o},\text{B},I)} \times \frac{\lambda_\text{max} - \lambda_\text{min}}{\delta\lambda \sqrt{2\pi}}. \quad (4.8)$$

The first term on the right-hand side reflects our relative prior preference for the alternative theories; to be fair, we can take it to be unity. The second term is a measure of how well the best predictions from each of the models agree with the data; with the added flexibility of his adjustable parameter, this maximum likelihood ratio can only favour B. The goodness-of-fit, however, cannot be the only thing that matters; if it was, we would always prefer more complicated explanations. Probability theory tells us that there is, indeed, another term to be considered. As assumed earlier in the evaluation of the marginal integral of eqn (4.3), the prior range $\lambda_\text{max} - \lambda_\text{min}$ will generally be much larger than the uncertainty $\pm\delta\lambda$ permitted by the data. As such, the final term in eqn (4.8) acts to penalize B for the additional parameter; for this reason, it is often called an *Ockham factor*. That is to say, we have naturally encompassed the spirit of Ockham's Razor: *'Frustra fit per plura quod potest fieri per pauciora'* or, in English, 'it is vain to do with more what can be done with fewer'.

Although it is satisfying to quantify the everyday guiding principle attributed to the thirteenth-century Franciscan monk William of Ockham (or *Occam*, in Latin), that we should prefer the simplest theory which agrees with the empirical evidence, we should not get too carried away by it. After all, what do we mean by the simpler theory if alternative models have the same number of adjustable parameters? In the choice between Gaussian and Lorentzian peak shapes, for example, both are defined by the position of the maximum and their width. All that we are obliged to do, and have done, in addressing such questions is to adhere to the rules of probability.

While accepting the clear logic leading to eqn (4.8), many people rightly worry about the question of the limits λ_min and λ_max. Jeffreys (1939) himself was concerned and pointed out that there would be an infinite penalty for any new parameter if the range was allowed to go to $\pm\infty$. Stated in the abstract, this would appear to be a severe limitation. In practice, however, it is not generally such a problem: since the analysis is always used in specific contexts, a suitable choice can usually be made on the basis of the relevant background information. Even in uncharted territory, a small amount of thought soon reveals that our state of ignorance is always far from the $\pm\infty$ scenario. If λ was the *coupling constant* (or strength) for a possible *fifth force*, for example, then we could put an upper bound on its magnitude because everybody would have noticed it by now if it had been large enough! We should also not lose sight of the fact that the precise form of eqn (4.8) stems from our stated simplifying approximations; if these are not appropriate, then eqns (4.2) and (4.3) will lead us to a somewhat different formula.

In most cases, our relative preference for A or B is dominated by the goodness of the fit to the data; that is to say, the maximum likelihood ratio in eqn (4.8) tends to overwhelm the contributions of the other two terms. The Ockham factor can play a crucial rôle, however, when both theories give comparably good agreement with the measurements. Indeed, it becomes increasingly important if B's theory fails to give a significantly better fit as the quality of the data improves. In that case, $\delta\lambda$ continues to become smaller but the ratio of best-fit likelihoods remains close to unity; according to eqn (4.8), therefore, A's theory is favoured ever more strongly. By the same token, the Ockham effect disappears if the data are either few in number, of poor quality or just fail to shed new light on the problem at hand. This is simply because the posterior ratio of eqn (4.1) is then roughly equal to the complementary prior one, since the empirical evidence is very weak; hence, there is no inherent preference for A's theory unless it is explicitly encoded in $\mathrm{prob}(A|I)/\mathrm{prob}(B|I)$. This property can be verified formally by going back to eqns (4.2), (4.5) and (4.6), and considering the poor-data limit in which $\delta\lambda \gg \lambda_{\max} - \lambda_{\min}$ and $\mathrm{prob}(D|\lambda_\mathrm{o}, B, I) \approx \mathrm{prob}(D|A, I)$.

Some further interesting features arise when we consider the case where Mr A also has one adjustable parameter; call it μ. If we make the same sort of probability assignments, and simplifying approximations, as for Mr B, then we find that

$$\frac{\mathrm{prob}(A|D, I)}{\mathrm{prob}(B|D, I)} = \frac{\mathrm{prob}(A|I)}{\mathrm{prob}(B|I)} \times \frac{\mathrm{prob}(D|\mu_\mathrm{o}, A, I)}{\mathrm{prob}(D|\lambda_\mathrm{o}, B, I)} \times \frac{\delta\mu\,(\lambda_{\max} - \lambda_{\min})}{\delta\lambda\,(\mu_{\max} - \mu_{\min})} . \quad (4.9)$$

This could represent the situation where we have to choose between a Gaussian and Lorentzian shape for a signal peak, but one associated parameter is not known. The position of the maximum may be fixed at the origin by theory, for example, and the amplitude constrained by the normalization of the data; A and B could then be the hypotheses favouring the alternative lineshapes, where μ and λ are their related full-width-half-maxima. If we give equal weight to A and B before the analysis, and assign a similar large prior range for both μ and λ, then eqn (4.9) reduces to

$$\frac{\mathrm{prob}(A|D, I)}{\mathrm{prob}(B|D, I)} \approx \frac{\mathrm{prob}(D|\mu_\mathrm{o}, A, I)}{\mathrm{prob}(D|\lambda_\mathrm{o}, B, I)} \times \frac{\delta\mu}{\delta\lambda} .$$

For data of good quality, the dominant factor will tend to be the best-fit likelihood ratio. If both give comparable agreement with the measurements, however, then the shape with the larger error-bar for its associated parameter will be favoured. At first sight, it might seem rather odd that the less discriminating theory can gain the upper hand. It appears less strange once we realize that, in the context of model selection, a larger 'error-bar' means that more parameter values are consistent with the given hypothesis; hence its preferential treatment.

Finally, we can also consider the situation where Mr A and Mr B have the same physical theory but assign a different prior range for λ (or μ). Although eqn (4.8) can be seen as representing the case when $\mu_{\max} - \mu_{\min}$ is infinitesimally small, so that A has no flexibility, eqn (4.9) is more appropriate when the limits set by both theorists are large enough to encompass all the parameter values giving a reasonable fit to the data. With equal initial weighting towards A and B, the latter reduces to

$$\frac{\mathrm{prob}(A|D, I)}{\mathrm{prob}(B|D, I)} = \frac{\lambda_{\max} - \lambda_{\min}}{\mu_{\max} - \mu_{\min}} ,$$

because the best-fit likelihood ratio will be unity (since $\lambda_o = \mu_o$) and $\delta\lambda = \delta\mu$. Thus, our analysis will lead us to prefer the theorist who gives the narrower prior range; this is not unreasonable as he must have had some additional insight to be able to predict the value of the parameter more accurately.

4.1.1 Comparison with parameter estimation

The dependence of the result in eqn (4.8) on the prior range $\lambda_{max} - \lambda_{min}$ can seem a little strange, since we haven't encountered such behaviour in the preceding chapters. It is instructive, therefore, to compare the model selection analysis with parameter estimation. To infer the value of λ from the data, given that B's theory is correct, we use Bayes' theorem:

$$\text{prob}(\lambda|D,\text{B},I) = \frac{\text{prob}(D|\lambda,\text{B},I) \times \text{prob}(\lambda|\text{B},I)}{\text{prob}(D|\text{B},I)} . \qquad (4.10)$$

The numerator is the familiar product of a prior and likelihood, and the denominator is usually omitted since it does not depend explicitly on λ; hence this relationship is often written as a proportionality. From the story of Mr A and Mr B, however, we find that the neglected term on the bottom plays a crucial rôle in ascertaining the merit of B's theory relative to a competing alternative. In recognition of its new-found importance, the denominator in Bayes' theorem is sometimes called the 'evidence' for B; it is also referred to as the 'marginal likelihood', the 'global likelihood' and the 'prior predictive'. Since all the components necessary for both parameter estimation and model selection appear in eqn (4.10), we are not dealing with any new principles; the only thing that sets them apart is that we are asking different questions of the data.

A simple way to think about the difference between parameter estimation and model selection is to note that, to a good approximation, the former requires the location of the maximum of the likelihood function whereas the latter entails the calculation of its average value. As long as λ_{min} and λ_{max} encompass the significant region of $\text{prob}(D|\lambda,\text{B},I)$ around λ_o, the precise bounds do not matter for estimating the optimal parameter and need not be specified. Since the prior range defines the domain over which the mean likelihood is computed, due thought is necessary when dealing with model selection. Indeed, it is precisely this act of comparing 'average' likelihoods rather than 'maximum' ones which introduces the desired Ockham balance to the goodness-of-fit criterion. Any likelihood gain from a better agreement with the data, allowed by the greater flexibility of a more complicated model, has to be weighed against the additional cost of averaging it over a larger parameter space.

It is important to remember that the discussion of the rôle of λ_{min} and λ_{max} in our illustrative example stems specifically from the assignment of eqn (4.4). In practice, a more suitable prior might be

$$\text{prob}(\lambda|\text{B},I) = \frac{e^{-\lambda/b}}{b} ,$$

where b is an initial estimate of the magnitude of the parameter λ, with $0 < \lambda < \infty$, but the probabilistic evidence for Mr. B's theory is still given by the prior-weighted average of his likelihood as in eqn (4.3). For clarity of exposition, however, we will continue to use eqn (4.4) and its multivariate generalization.

4.1.2   Hypothesis testing

The phrasing of the story of Mr A and Mr B suggests that we are dealing with the problem conventionally called *hypothesis testing*. Although this is usually treated as a separate topic in the orthodox literature, our purpose is to emphasize that the principles required are no different from those used in parameter estimation: it's all just a matter of applying the sum and product rules of probability theory.

Suppose we have a hypothesis H_1: the shape of the signal peak is Gaussian, for example. To quantify how much we believe that this proposition is true, based on the data D and all the relevant background information I, we need to evaluate the posterior probability $\mathrm{prob}(H_1|D, I)$; to help us do this, we can use Bayes' theorem:

$$\mathrm{prob}(H_1|D, I) = \frac{\mathrm{prob}(D|H_1, I) \times \mathrm{prob}(H_1|I)}{\mathrm{prob}(D|I)} \,. \qquad (4.11)$$

The numerator is simply the product of the prior and the likelihood for H_1, but we also require $\mathrm{prob}(D|I)$ to put the posterior probability on a meaningful scale. The denominator can be ignored, of course, if we are only interested in the relative merit of H_1 compared to another hypothesis H_2; then, applying Bayes' theorem to the second proposition, and dividing by the expression above, we obtain the odds-ratio:

$$\frac{\mathrm{prob}(H_1|D, I)}{\mathrm{prob}(H_2|D, I)} = \frac{\mathrm{prob}(D|H_1, I)}{\mathrm{prob}(D|H_2, I)} \times \frac{\mathrm{prob}(H_1|I)}{\mathrm{prob}(H_2|I)} \,,$$

which is the same as eqn (4.2), with $H_1 = \mathrm{A}$ and $H_2 = \mathrm{B}$. To assess the intrinsic truth of H_1, we could think of letting H_2 be the hypothesis that H_1 is false: $H_2 = \overline{H_1}$. Using marginalization and the product rule, we would then write $\mathrm{prob}(D|I)$ as

$$\mathrm{prob}(D|I) = \mathrm{prob}(D|H_1, I) \times \mathrm{prob}(H_1|I) + \mathrm{prob}(D|\overline{H_1}, I) \times \mathrm{prob}(\overline{H_1}|I) \,,$$

where the two priors would be related by the sum rule of eqn (1.1):

$$\mathrm{prob}(H_1|I) + \mathrm{prob}(\overline{H_1}|I) = 1 \,.$$

In conjunction with eqn (4.11), we see that the difficult term is the likelihood function for $\overline{H_1}$: $\mathrm{prob}(D|\overline{H_1}, I)$. This is because, in general, we cannot compare the predictions of the hypothesis with the data given only that H_1 is false: we need well-defined alternatives! If the peak shape is not Gaussian, for example, could it be Lorentzian? With a specific set of possibilities, $\{H_j\}$, the problem boils down to one of model selection — simply compare their evidences, $\mathrm{prob}(D|H_j, I)$.

It is often said that a hypothesis should be *rejected* if it gives a poor fit to the data, even though agreement does not assure its truth. As such, traditional hypothesis testing often involves the use of procedures designed to asses the *significance* of the mismatch between theory and experiment; many entail the χ^2 statistic. As shown by eqn (3.83), it is certainly true that if we know the nature of the object of interest, and the related data are subject to independent Gaussian noise, then the expected value of χ^2 will be about equal to the number of measurements N; we would be quite surprised by deviations of more than a few times $\sqrt{N}$. Nevertheless, it is a bold step to go from this statement to rejecting a hypothesis because χ^2 is too large. The point is that the misfit statistic is a measure of the likelihood function $\text{prob}(data \,|\, hypothesis, I)$; to reject (or accept) a theory, however, we need the posterior probability $\text{prob}(hypothesis \,|\, data, I)$. Although a larger value of χ^2 will give a smaller likelihood for the data, it is only one of the ingredients which makes up the posterior: we also need $\text{prob}(data \,|\, I)$, as well as an assignment for the prior. Despite the conventional practice of quoting *P-values* for assessing the level of rejection of a hypothesis, the basis for this formal quantification seems rather doubtful. The misfit statistic can serve a useful (qualitative) purpose, nonetheless, if a poor quality-of-fit prompts us to think about alternative hypotheses; probability theory then provides us with the tools for quantitatively choosing which one is best.

Even if we could come up with a convincing argument to disregard a theory, the job is only half-done until we have something better to take its place. To quote Jeffreys (1939, Section 7.2.2), '*If there is no clearly stated alternative, and the null hypothesis is rejected, we are left without any rule at all, whereas the null hypothesis, though not satisfactory, may at any rate show some sort of correspondence with the facts.*' He points out that, while there was never a time when Newton's theory of gravity would not have failed a *P*-test, 'The success of Newton was not that he explained all the variation of the observed positions of the planets, but that he explained most of it.'

4.2 Example 6: how many lines are there?

To illustrate the practical use of the analysis discussed above, let us consider a case which occurs frequently in spectroscopy, crystallography and many other areas of science: namely, the question of assessing how many signal peaks there is most evidence for in a pertinent set of data. Here we are primarily concerned with inferring the amplitudes and positions of a set of a *few* discrete excitation lines, but are not sure as to their exact number.

For simplicity, let's assume that we are dealing with a one-dimensional problem where the shape of the signal peaks $f(x)$ is known; then, the ideal spectrum $G(x)$ can be expressed as:

$$G(x) = \sum_{j=1}^{M} A_j \, f(x, x_j) \,, \tag{4.12}$$

where A_j is the magnitude of the jth line and x_j represents its location. If all the excitations were Gaussian with width W, for example, then

$$f(x, x_j) = \exp\left[-\frac{(x - x_j)^2}{2\,W^2} \right].$$

The situation is shown in Fig. 4.2(a). Most experimental set-ups give rise to a blurred and noisy version of the spectrum in eqn (4.12), and tend to entail a slowly varying

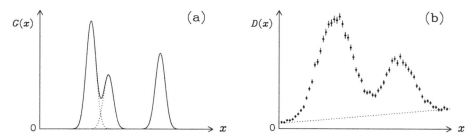

Fig. 4.2 (a) An ideal spectrum consisting of the sum of a few signal peaks; in the model selection problem, we're not sure as to their number. (b) The corresponding data are usually a blurred and noisy version of (a), and are often corrupted by a slowly varying background (dotted line).

background signal $B(x)$; such measurements are illustrated in Fig. 4.2(b). In this case, the ideal data $\{F_k\}$ are related to the parameters $\{A_j, x_j\}$ by

$$F_k = \int G(x)\, R(x_k - x)\, \mathrm{d}x \;+\; B(x_k)\,, \tag{4.13}$$

where we have implicitly assumed that the shape of the *resolution function* $R(x)$ does not vary with position in writing the blurring process as a convolution integral. If we make the approximation that the experimental measurements $\{D_k\}$ are subject to independent additive Gaussian noise $\{\sigma_k\}$, then we obtain the least-squares likelihood function of Section 3.5:

$$\mathrm{prob}(\{D_k\}|\{A_j, x_j\}, M, I) \;\propto\; \exp\!\left(-\frac{\chi^2}{2}\right), \tag{4.14}$$

where χ^2 is defined in eqn (3.66) and the normalization factor, involving the product of terms like $\sigma_k \sqrt{2\pi}$, has been omitted since its value is independent of the parameters to be inferred. Note that we have stated the conditioning on the number of excitation lines M explicitly, whereas R, B and $\{\sigma_k\}$ have been subsumed in I. This is because its value is not known, but is of interest to us: hence, it's a model selection problem.

To estimate the number of signal peaks, we need to evaluate the posterior pdf for M, $\mathrm{prob}(M|\{D_k\}, I)$. As always, it's best to start with Bayes' theorem:

$$\mathrm{prob}(M|\{D_k\}, I) \;=\; \frac{\mathrm{prob}(\{D_k\}|M, I) \times \mathrm{prob}(M|I)}{\mathrm{prob}(\{D_k\}|I)}\,.$$

If we assign a simple uniform prior for $M = 1, 2, \ldots$, up to a few (say 20), then the posterior becomes proportional to the evidence (or marginal likelihood):

$$\mathrm{prob}(M|\{D_k\}, I) \;\propto\; \mathrm{prob}(\{D_k\}|M, I)\,, \tag{4.15}$$

where the omitted normalization constant can be evaluated at the end, if required, from the condition $\sum \mathrm{prob}(M|\{D_k\}, I) = 1$. As in the case of Mr B in the previous section, the evidence can be expressed as a marginal integral over the product of the prior and likelihood function for the parameters of the M-line model:

$$\text{prob}(\{D_k\}|M,I) = \iint \cdots \int \text{prob}(\{D_k\},\{A_j,x_j\}|M,I)\, \text{d}^M A_j\, \text{d}^M x_j\, , \quad (4.16)$$

where

$$\text{prob}(\{D_k\},\{A_j,x_j\}|M,I) = \text{prob}(\{D_k\}|\{A_j,x_j\},M,I)\, \text{prob}(\{A_j,x_j\}|M,I)\, .$$

Since we've already settled on a least-squares likelihood in eqn (4.14), all we need now is an assignment for the prior. In keeping with our policy of naïveté, let us take this to be a simple uniform pdf within a suitably bounded region:

$$x_{\text{min}} \leqslant x_j \leqslant x_{\text{max}} \quad \text{and} \quad 0 \leqslant A_j \leqslant A_{\text{max}}\, , \quad (4.17)$$

and zero otherwise. This means that, to be properly normalized, the prior is given by the reciprocal of the volume of the hypercube defined by eqn (4.17):

$$\text{prob}(\{A_j,x_j\}|M,I) = \left[(x_{\text{max}} - x_{\text{min}})\, A_{\text{max}}\right]^{-M}\, , \quad (4.18)$$

and is just a multidimensional generalization of Mr B's prior for λ in eqn (4.4). The parameters x_{min} and x_{max} can be set equal to the x-limits over which the data have been measured; if we had thought that the lines of interest could lie outside this range, we would presumably have conducted a different experiment. If we were not already given it, we could take an upper bound for the amplitudes of the peaks A_{max} from the integrated intensity of the data; we'd effectively be making use of the known product of the overall experimental count-rate and the length of time for which the measurements were made.

Substituting the product of the likelihood and prior of eqns (4.14) and (4.18) into eqn (4.16), we see that the posterior in eqn (4.15) is given by

$$\text{prob}(M|\{D_k\},I) \propto \left[(x_{\text{max}} - x_{\text{min}})\, A_{\text{max}}\right]^{-M} \iint \cdots \int \exp\left(-\frac{\chi^2}{2}\right) \text{d}^M A_j\, \text{d}^M x_j\, ,$$
$$(4.19)$$

where the multiple integral is over the region defined by eqn (4.17). Since the pdf $\exp(-\chi^2/2)$ can have a very complicated multimodal topology, it is safest to carry out this integration numerically with a Monte Carlo algorithm. Alternatively, we can pursue the calculation further analytically by making some gross simplifying approximations. Suppose that there is a set of $2M$ parameters $\boldsymbol{X}_{\text{O}} = \{A_{\text{o}j}, x_{\text{o}j}\}$ which yield the best least-squares fit to the data χ^2_{min}; then, a quadratic Taylor series expansion about this point gives

$$\chi^2 \approx \chi^2_{\text{min}} + \tfrac{1}{2}\left(\boldsymbol{X} - \boldsymbol{X}_{\text{O}}\right)^{\text{T}} \boldsymbol{\nabla}\boldsymbol{\nabla}\chi^2(\boldsymbol{X}_{\text{O}})\left(\boldsymbol{X} - \boldsymbol{X}_{\text{O}}\right) + \cdots\, ,$$

as in eqn (3.30). Assuming that the significant portion of the pdf $\exp(-\chi^2/2)$, around the maximum at $\boldsymbol{X}_{\text{O}}$, lies within the region permitted by eqn (4.17), the contribution of this solution towards eqn (4.19) is equal to $\exp(-\chi^2_{\min}/2)$ times the integral of the related $2M$-dimensional multivariate Gaussian:

$$\int\!\!\int \cdots \int \exp\left[-\tfrac{1}{4}(X-X_0)^T \nabla\nabla\chi^2(X_0)(X-X_0)\right] d^{2M}X_j = \frac{(4\pi)^M}{\sqrt{\det(\nabla\nabla\chi^2)}},$$

where $\det(\nabla\nabla\chi^2)$ is the determinant of the Hessian matrix, evaluated at X_0. As the labelling of the M lines in our model of the spectrum is arbitrary (which one we call 3 and which one 4 etc.), there will be $M!$ equivalent maxima in the likelihood function; this is just the number of ways we can *permute* the indices associated with the various signal peaks. Hence, we can approximate the posterior pdf of eqn (4.19) by

$$\text{prob}(M|\{D_k\},I) \propto \frac{M!\,(4\pi)^M}{\left[(x_{\max}-x_{\min})A_{\max}\right]^M \sqrt{\det(\nabla\nabla\chi^2)}} \exp\left(-\frac{\chi^2_{\min}}{2}\right).$$

(4.20)

As mentioned earlier, the omitted constant of proportionality can be determined from the normalization requirement: $\sum \text{prob}(M|\{D_k\},I) = 1$.

Although eqn (4.20) looks rather horrendous, there is a close correspondence with the decomposition of eqn (4.8) in the analysis of Mr A and Mr B. The prior term $\text{prob}(M|I)$ is missing because we originally had no (strong) preference for any particular number of lines; therefore, it has been absorbed into the normalization constant. With the assumption that the data are subject to independent Gaussian noise, the best-fit likelihood is proportional to $\exp(-\chi^2_{\min}/2)$. The remaining terms constitute the multivariate equivalent of the Ockham factor: they are the ratio of the $2M$-dimensional hypervolumes, in the parameter space $\{A_j, x_j\}$, permitted by the prior and the posterior. We should emphasize that the specific form of eqn (4.20) relies on the validity of our simplifying approximations. They will tend to be reasonable in our stated (prior) régime of small M, but we must return to a more robust evaluation of the multiple integral in eqn (4.19) if they are not adequate. Indeed, if the assignment of the least-squares likelihood, or the uniform prior, is not appropriate, then we have to take a step further back to eqn (4.16).

In the above analysis, we have implicitly assumed (in I) that the exact shape of the signal peak, and the background, is known. This is often the case, as they are usually determined fairly accurately from calibration experiments. If we are not quite so fortunate, however, then we must characterize their nature by a small number of parameters and marginalize over them. For example, a slowly varying background can be approximated by a straight line if the x-range of the data is not too large; $B(x)$ is, therefore, defined by just two unknown variables b_1 and b_2. The conditioning of the pdfs on these parameters must now be stated explicitly, so that eqn (4.16) becomes

$$\text{prob}(\{D_k\}|M,I) = \int\!\!\int \cdots \int \text{prob}(\{D_k\}, \{A_j, x_j\}, b_1, b_2|M,I)\, d^M A_j\, d^M x_j\, db_1\, db_2.$$

Proceeding as before, we will be led to the formula in eqn (4.20); the only difference is that b_1 and b_2 will affect both the best-fit possible $\chi^2_{\min}$ and contribute towards the determinant of the $\nabla\nabla\chi^2$ matrix (now $2M+2$ square). Since the prior pdf for b_1 and b_2 does not depend on the number of lines, $\mathrm{prob}(b_1, b_2 | M, I) = \mathrm{prob}(b_1, b_2 | I)$, its normalization factor can be incorporated into the existing constant of proportionality.

Uncertainty about the width of excitation lines, or the size of the error-bars for the data, or any other systematic effect, can be dealt with in the same way through marginalization; the relevant integrations can be done by optimization, within the quadratic approximation. If there is debate as to whether the shape of the signal peaks is Gaussian (G) or Lorentzian (L), for example, then we can compare their evidence:

$$\frac{\mathrm{prob}(\{D_k\} | G, I)}{\mathrm{prob}(\{D_k\} | L, I)} = \frac{\int \mathrm{prob}(\{D_k\} | W, G, I)\, \mathrm{prob}(W | G, I)\, \mathrm{d}W}{\int \mathrm{prob}(\{D_k\} | W, L, I)\, \mathrm{prob}(W | L, I)\, \mathrm{d}W} ,$$

where we have assumed that all the lines have the same, but unknown, width W. If we take the priors $\mathrm{prob}(W | G, I)$ and $\mathrm{prob}(W | L, I)$ to be uniform over a comparable range ($W_{\min}$ to $W_{\max}$), then these terms will cancel. The remaining pdfs on the right can be expressed as marginal distributions over the number of lines:

$$\mathrm{prob}(\{D_k\} | W, G, I) = \sum_M \mathrm{prob}(\{D_k\} | M, W, G, I)\, \mathrm{prob}(M | W, G, I) ,$$

and there will be an identical equation with L's instead of G's. Since we don't expect a knowledge of the line shape to influence what we can say about M before the analysis of the data, we can assign $\mathrm{prob}(M | W, G, I) = \mathrm{prob}(M | W, L, I) = \mathrm{prob}(M | I)$. As the order of an integration and summation can be interchanged, we find that the required probability is given by

$$\mathrm{prob}(\{D_k\} | G, I) = \sum_{M=1}^{\mathrm{few}} \int_{W_{\min}}^{W_{\max}} \mathrm{prob}(\{D_k\} | M, W, G, I)\, \mathrm{d}W ,$$

where the integrand is just the evidence for the number of lines calculated earlier, when we were given the exact shape of the peaks. Additionally, we could evaluate the evidence for M by marginalizing over the two candidate line shapes:

$$\begin{aligned}
\mathrm{prob}(\{D_k\} | M, I) = {} & \mathrm{prob}(\{D_k\} | G, M, I) \times \mathrm{prob}(G | M, I) \\
& + \mathrm{prob}(\{D_k\} | L, M, I) \times \mathrm{prob}(L | M, I) ,
\end{aligned}$$

and so on.

4.2.1 An algorithm

In order to make use of the preceding analysis, we need an algorithm for its practical implementation. To this end, we will work with the approximation of eqn (4.20). This formula for calculating the posterior probability for the number of peaks requires us to find the value of $\chi^2_{\min}$, and the determinant of $\nabla\nabla\chi^2$ (evaluated at that optimal point in parameter space), for an M-line model. Both quantities should be readily available in standard least-squares programs because the minimum value of χ^2 defines the best-fit parameters and, from eqn (3.71), twice the inverse of $\nabla\nabla\chi^2$ gives an estimate of their reliability. Thus we have another example where a basic understanding of probability theory allows us to extract more potency from a widely-used procedure than a cookbook approach to least-squares would have suggested was possible. The main difference is that we must now give some thought to a suitable range for the uniform prior for the model variables, whereas previously it had been sufficient to just say that it was large.

The most difficult task for using eqn (4.20) is finding the set of parameters $\{A_j, x_j\}$ (and possibly b_1, b_2 and W) which yield the best-fit χ^2_{min}. This is because the positions of the peaks, unlike their amplitudes, are not linearly related to the data and so χ^2 can be multimodal; as discussed in Section 3.4, it leads to a (potentially) hard problem in multidimensional optimization. Most least-squares programs avoid this difficulty by shifting the onus on to the user to provide an initial guess which is good enough to refine to (what is hopefully) the global minimum. As a somewhat ambitious alternative, we outline below an algorithm based on the combination of one-dimensional 'brute force and ignorance' searches and linearized multidimensional optimization. Although far from being foolproof, it's computationally efficient and has been used with a fair amount of success.

Let us start by considering the range parameters x_{min}, x_{max} and A_{max}. Ideally, these should be determined from previous measurements, or theory, since they reflect our background information; in practice, a reasonable estimate can be obtained from a cursory look at the current data. As mentioned earlier, x_{min} and x_{max} can be set equal to the limits of the x-range over which the data have been measured. If the candidate signal peaks $f(x)$ in eqn (4.12), and the resolution function $R(x)$ in eqn (4.13), have been normalized with respect to x, then the mathematical properties of a convolution ensure that the integrated intensity of the data provides an upper bound for the amplitudes: $A_{\text{max}} \approx \sum D_k$.

To find the best-fit solution for any specified number of lines, let's begin with the case $M = 0$ and work upwards. If there are no signal peaks, then the data are simply described by a background $B(x)$. Assuming that this varies sufficiently slowly, it can be taken to be linear over the range x_{min} to x_{max}: $B(x) = b_1 x + b_2$. It is then easy to refine b_1 and b_2 to obtain both χ^2_{min} and the determinant of the 2×2 $\nabla\nabla\chi^2$ matrix; $\text{prob}(M = 0 | \{D_k\}, I)$ is just proportional to $\exp(-\chi^2_{\text{min}}/2)/\sqrt{\det(\nabla\nabla\chi^2)}$. Next, we need the best-fit parameters for a one-line model. To circumvent the problem of non-linearity associated with the position x_1, we can conduct an explicit search for its value: that is, we divide the x-range into a couple of hundred discrete points and optimize the linear parameters A_1, b_1 and b_2 as we step through the finite set of possibilities for x_1. Taking the values which yield the smallest χ^2 as a good initial guess, we can carry out a four-dimensional search for the best-fit solution for the one-line model. To improve robustness, we can use a simplex routine before completing the refinement with Newton–Raphson. The former has the additional advantage that, since it works directly with χ^2, it allows for the encoding of the sharp bounds of the uniform prior in eqn (4.17); nevertheless, the efficiency of Newton–Raphson ensures that it is a useful last step. In any case, we need the gradient information in $\nabla\nabla\chi^2$ to estimate both the posterior probability for the one-line model and the error-bars for its parameters; since analytical differentiation with respect to x_1 is awkward, finite differences can be used to obtain the derivatives.

The rest of the algorithm is essentially a repetition of this recipe. For the second line, we first carry out an explicit search for x_2 on a one-dimensional grid; during this procedure, we can refine all the linear parameters (such as A_1, A_2, b_1 and b_2) but it is best to hold the others (like x_1) fixed at their previous values. This is then followed by a simultaneous optimization of all the parameters of the two-line model, with simplex and Newton–Raphson. Having evaluated χ^2 and $\nabla\nabla\chi^2$ at the optimal solution, we can calculate the posterior $\mathrm{prob}(M = 2\,|\,\{D_k\}, I)$ from eqn (4.20). This process of combining one-dimensional brute force and ignorance searches with linearized multidimensional refinement can be continued until we reach the upper limit of our prior $\mathrm{prob}(M\,|\,I)$.

Before illustrating this algorithm with a couple of examples, we should make some additional remarks. It is helpful to use the prior ranges $A_{\max}$ and $x_{\max}-x_{\min}$ (and even $b_{\max}$, for the background) to scale the parameters to be optimized. This is because then all the variables, such as $A_j/A_{\max}$, will be dimensionless and of comparable magnitude. Another advantage of working in this scaled space is that we can improve the stability of the matrix calculations by adding a small multiple of the identity matrix to the Hessian: $\nabla\nabla\chi^2 \to \nabla\nabla\chi^2 + c\,\mathbf{I}$. As shown in eqn (3.58), this has no effect on the eigenvectors of the Hessian and so does not change the pattern of correlations between the inferred parameters. It does, however, put a lower bound on the eigenvalues ($\geqslant c$), thereby encoding our prior knowledge that the error-bar for any parameter cannot be larger than the prior range assigned to it. Finally, if the width W of the peaks is not known then it too must be marginalized like the background. Because of its highly non-linear nature, it is often best done by running the program several times over using different given widths. We can then plot the two-dimensional posterior pdf for the number of lines and their width, and integrate with respect to W to obtain $\mathrm{prob}(M\,|\,\{D_k\}, I)$. If desired, we can sum over M to yield $\mathrm{prob}(W\,|\,\{D_k\}, I)$.

4.2.2 Simulated data

Let us begin our illustration of model selection with the aid of data generated in a computer simulation. The measurements are shown in Fig. 4.3(a) and result from the convolution of a spectrum consisting of the sum of a 'few' excitation lines, all having the same Gaussian profile, with a Gaussian resolution function of FWHM 2.0 μeV; they are also corrupted by a linear background and (Poisson) noise. Given this information alone, how many lines is there most evidence for in the data?

Carrying out the analysis with the algorithm described above, we obtain the posterior pdf for M indicated by the triangles in Fig. 4.3(b). Note that, to highlight its shape, the pdf has been plotted on a logarithmic scale and a continuous solid line drawn between the discrete points. The position of the maximum indicates that there is most evidence for two signal peaks. Their width W (FWHM) is estimated to be $1.03 \pm 0.08\,\mu$eV, with locations at $13.98 \pm 0.03\,\mu$eV and $15.47 \pm 0.02\,\mu$eV. The spectrum used to generate the data in Fig. 4.3(a) did indeed contain two lines, with a FWHM of $1.0\,\mu$eV, centred at 14.0 and $15.5\,\mu$eV. The amplitudes of both were the same and were inferred correctly to within 5%.

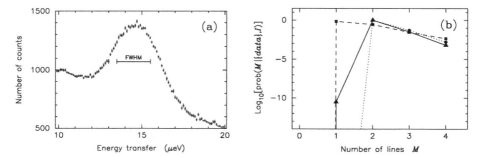

Fig. 4.3 (a) Simulated data from a spectrum of a few excitation lines; the width of the instrumental resolution function is $2.0\,\mu\text{eV}$. (b) The triangles, and solid line, mark the corresponding posterior probability for the number of lines M. The dashed line shows the result for data of poorer quality, with this shortfall supplemented by increased prior knowledge in the analysis giving the dotted line.

The shape of the posterior pdf for M is characteristic of this type of model selection analysis: (i) there is a sharp fall-off from the maximum on the left, because there is not enough structure in the proposed spectrum to adequately account for the data; (ii) there is a slow decline on the right, as the models become increasingly, and unnecessarily, complicated. As discussed in Section 4.1, we have captured the essence of Ockham's Razor: our best estimate of the spectrum is the one with the least number of lines that is consistent with the experimental measurements. Although this is exactly what we would have done using our 'common sense', the value of the analysis is that it sharpens and refines it far beyond our qualitative intuition. It would otherwise be difficult for us to state that the two-line model was ten orders of magnitude more probable than the one-line alternative, or that the empirical evidence for two peaks was thirty times greater than that for three.

We should emphasize, however, that the conclusions are always conditional on both the quality of the data and our prior knowledge. For example, if the error-bars for the data of Fig. 4.3(a) had been three times larger (e.g. experiment run for only one-tenth of the time), we would have obtained the posterior pdf marked by the squares, and dashed line, in Fig. 4.3(b). There is then most evidence for only one line, at $\sim 14.8\,\mu\text{eV}$, with a FWHM of about $2\,\mu\text{eV}$, although two could not be ruled out at the 90% confidence level. This is because the poorer measurements can be more simply, but sufficiently, explained by a broader single peak than with two narrower ones. Indeed, the maximum of the posterior would occur at $M = 0$ if the error-bars were so large that the data could be fit adequately by just a linear background! In that case the pdf $\text{prob}(M\,|\,\{data\}, I)$ would also be very flat, closely resembling the prior $\text{prob}(M\,|\,I)$; consequently, probability theory would be warning us that it was unwise to make too decisive a judgement based on such flimsy measurements. Even with the poorer data, with three times more noise than in Fig. 4.3(a), we would still find most evidence for two lines if it was already known that the peaks comprising the spectrum had a FWHM of $1.0\,\mu\text{eV}$; the corresponding posterior is shown by the circles, and dotted line, in Fig. 4.3(b).

4.2.3 Real data

An interesting series of illustrations of the (real-life) use of the theory and algorithm outlined above, as applied to molecular tunnelling spectroscopy with neutron scattering data, can be found in Sivia and Carlile (1992). Here we present just one example from crystallography: the analysis of part of the X-ray diffraction pattern from a powdered sample of a zeolite. The experimental data were collected at the synchrotron source at Daresbury, U.K., and are shown in Fig. 4.4(a). The shape of the Bragg peaks, in Fig. 4.4(b), is known from calibration measurements; it does not vary with scattering angle θ. It is the effective resolution function $R(\theta)$ for the experiment and includes the contribution from both the intrinsic width of the lines and the instrumental blurring.

The first step in solving the structure of a crystal involves the determination of the size of the *unit cell* and its *space-group* symmetry. To tackle both these questions, we need to estimate the positions and amplitudes of the Bragg lines; before we can do that, however, we need to know how many there are. Carrying out the analysis as described earlier, we obtain the posterior pdf for the number of Bragg peaks given in Fig. 4.4(c). There is most evidence for five lines, therefore, and the optimal estimates of their amplitudes $\{A_j\}$ and positions $\{\theta_j\}$ are shown in Fig. 4.4(d); the (1-σ) error-bars for these parameters are also indicated, but the horizontal discrepancies for the locations are too

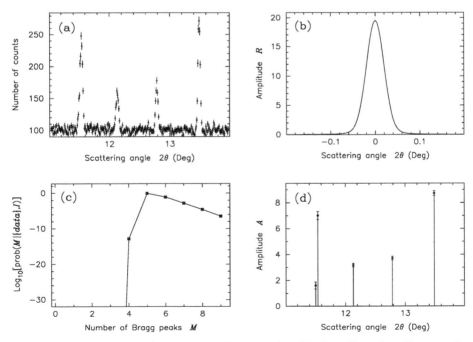

Fig. 4.4 (a) Part of the X-ray diffraction data from a zeolite. (b) The calibrated profile, or resolution function, of the Bragg peaks. (c) The logarithm of the posterior pdf for the number of lines. (d) The inferred amplitudes and positions of the Bragg peaks, and their estimated error-bars.

small to be seen. The ability to infer, with a high degree of confidence, that one of the visible peaks is a doublet, whereas the others are single, can be crucial to the correct assignment of the space-group and thence the successful solution of the crystal structure.

4.3   Other examples: means, variance, dating and so on

The model selection procedure described in this chapter is applicable to many problems in data analysis. Bretthorst (1988, 1990), for example, has been using it in NMR spectroscopy; atmospheric variations of ^{14}C were analysed by Sonett (1990); and Gull (1988) has addressed the perennial question of the optimal expansion-order (of a polynomial function) for fitting graphical data. Let us conclude this discussion with some further illustrations.

4.3.1   The analysis of means and variance

A common problem in science is one of *classification*. An archaeologist, for example, might have found two sets of humanoid skeletons whose ages differ by a million years. A question which could arise is whether variations in physical features, such as their brain size, show evidence for significant evolutionary change. Suppose that the two sites yield N_1 and N_2 measurements, represented by data-vectors D_1 and D_2, respectively; we could then consider the relative merit of the following hypotheses:

A: there is no change over this period of time, and so both sets of data can be characterized by the same (unknown) mean μ and standard deviation σ.
B: there is a change, with the two sites having (unknown) means μ_1 and μ_2, and standard deviations σ_1 and σ_2, respectively.

In principle, there is also the possibility that the mean remains the same while the standard deviation changes and *vice versa*; unless we have good reason to expect such peculiar one-sided evolution, however, we would tend to assign a fairly low prior probability to these propositions. Thus, it will generally be adequate to consider only the case of A and B above; a more complete analysis, and an outline of the history of the problem, can be found in Bretthorst (1993). Essentially, we are saying that our task here is one of deciding whether all the skulls belong to one class or two distinctly different ones.

In Section 4.1, we saw that the data-dependent term in model selection was the evidence; therefore, we need to evaluate $\mathrm{prob}(D_1, D_2 | \mathrm{A}, I)$ and $\mathrm{prob}(D_1, D_2 | \mathrm{B}, I)$. Beginning with the former, marginalization and the product rule allow us to write

$$\mathrm{prob}(D_1, D_2 | \mathrm{A}, I) = \iint \mathrm{prob}(D_1, D_2 | \mu, \sigma, \mathrm{A}, I)\, \mathrm{prob}(\mu, \sigma | \mathrm{A}, I)\, \mathrm{d}\mu\, \mathrm{d}\sigma, \quad (4.21)$$

For the prior for A's parameters, we can assign a simple uniform pdf in the region $\mu_{\min} \leqslant \mu \leqslant \mu_{\max}$ and $0 \leqslant \sigma \leqslant \sigma_{\max}$,

$$\mathrm{prob}(\mu, \sigma | \mathrm{A}, I) = \left[(\mu_{\max} - \mu_{\min})\, \sigma_{\max} \right]^{-1}, \quad (4.22)$$

and zero otherwise; we could even let $\mu_{\min} = 0$ and $\mu_{\max} = \sigma_{\max}$, and set them equal to the maximum extent of the measuring equipment. Since we believe that the variations in the brain size of the specimens can be characterized by just a mean and a standard deviation, it is reasonable to assign a Gaussian pdf for the likelihood function; we will formally see why this is so in Chapter 5. Treating the data as a single set of $N = N_1 + N_2$ independent measurements $\{x_k\}$, as in eqn (3.35), we have

$$\text{prob}(\boldsymbol{D}_1, \boldsymbol{D}_2 | \mu, \sigma, A, I) = \left(\sigma\sqrt{2\pi}\right)^{-N} \exp\left[-\frac{1}{2\sigma^2} \sum_{k=1}^{N} (x_k - \mu)^2\right]. \qquad (4.23)$$

The double integral of eqn (4.21) can be calculated numerically, by evaluating the pdf of eqn (4.23) on a rectangular grid defined by the prior of eqn (4.22) and summing the contributions. It can also be approximated analytically, with a quadratic Taylor series expansion of the logarithm of A's likelihood function:

$$L = L(\mu_o, \sigma_o) - \frac{1}{2} \begin{pmatrix} \mu - \mu_o & \sigma - \sigma_o \end{pmatrix} \begin{pmatrix} \alpha & \gamma \\ \gamma & \beta \end{pmatrix} \begin{pmatrix} \mu - \mu_o \\ \sigma - \sigma_o \end{pmatrix} + \cdots, \qquad (4.24)$$

where $L = \log_e[\text{prob}(\boldsymbol{D}_1, \boldsymbol{D}_2 | \mu, \sigma, A, I)]$, with a maximum at (μ_o, σ_o). Accordingly, the parameters μ_o and σ_o are determined by the conditions $\partial L/\partial \mu = 0$ and $\partial L/\partial \sigma = 0$:

$$\mu_o = \frac{1}{N} \sum_{k=1}^{N} x_k \quad \text{and} \quad \sigma_o^2 = \frac{1}{N} \sum_{k=1}^{N} (x_k - \mu_o)^2, \qquad (4.25)$$

and the elements of the 2×2 matrix are given by the second partial derivatives of L, evaluated at μ_o and σ_o:

$$\alpha = N/\sigma_o^2, \qquad \beta = 2N/\sigma_o^2, \qquad \gamma = 0. \qquad (4.26)$$

Exponentiating the log-likelihood of eqn (4.24), and using the prior of eqn (4.22), we see that eqn (4.21) can be approximated by the product of

$$\frac{\exp\left[L(\mu_o, \sigma_o)\right]}{(\mu_{\max} - \mu_{\min})\,\sigma_{\max}} = \frac{\text{prob}(\boldsymbol{D}_1, \boldsymbol{D}_2 | \mu_o, \sigma_o, A, I)}{(\mu_{\max} - \mu_{\min})\,\sigma_{\max}}, \qquad (4.27)$$

and the integral of a bivariate Gaussian:

$$\int_{\mu_{\min}}^{\mu_{\max}} \int_{0}^{\sigma_{\max}} \exp\left(-\frac{1}{2}\left[\alpha\,(\mu - \mu_o)^2 + \beta\,(\sigma - \sigma_o)^2\right]\right) d\mu\,d\sigma \approx \frac{2\pi}{\sqrt{\alpha\,\beta}}. \qquad (4.28)$$

Substituting for the parameters α, β, μ_o and σ_o from eqns (4.25) and (4.26) into eqns (4.23), (4.27) and (4.28), the evidence for hypothesis A reduces to

$$\text{prob}(\boldsymbol{D}_1, \boldsymbol{D}_2 | A, I) \approx \frac{\left(\sigma_o\sqrt{2\pi}\right)^{2-N} \exp(-N/2)}{(\mu_{\max} - \mu_{\min})\,\sigma_{\max}\,N\sqrt{2}}. \qquad (4.29)$$

The other quantity we need for our classification problem is $\mathrm{prob}(D_1, D_2 | \mathrm{B}, I)$. Since hypothesis B asserts that the data-sets D_1 and D_2 pertain to distinctly different classes, knowledge of one tells us nothing about the other; in conjunction with the product rule of probability, this independence takes the form

$$\mathrm{prob}(D_1, D_2 | \mathrm{B}, I) = \mathrm{prob}(D_1 | \mathrm{B}, I) \times \mathrm{prob}(D_2 | \mathrm{B}, I) . \qquad (4.30)$$

Both terms on the right-hand side can be written as marginal integrals like eqn (4.21):

$$\mathrm{prob}(D_j | \mathrm{B}, I) = \iint \mathrm{prob}(D_j | \mu_j, \sigma_j, \mathrm{B}, I) \, \mathrm{prob}(\mu_j, \sigma_j | \mathrm{B}, I) \, \mathrm{d}\mu_j \, \mathrm{d}\sigma_j ,$$

where $j = 1, 2$. The priors for the two means and standard deviations can each be set equal to eqn (4.22), since we suspect they might even be the same; the likelihoods for D_1 and D_2 will be similar to eqn (4.23), with the relevant summation over the respective N_1 and N_2 data. A quadratic Taylor series expansion then allows us to approximate the probabilities on the right of eqn (4.30) by expressions like eqn (4.29):

$$\mathrm{prob}(D_j | \mathrm{B}, I) \approx \frac{\left(\sigma_{oj} \sqrt{2\pi}\,\right)^{2-N_j} \exp(-N_j/2)}{(\mu_{\max} - \mu_{\min}) \, \sigma_{\max} \, N_j \sqrt{2}} ,$$

where σ_{o1} and σ_{o2} are given by the appropriate summations in eqn (4.25), over D_1 and D_2 respectively. Finally, dividing eqn (4.29) by the resultant product for the evidence for hypothesis B, we obtain

$$\frac{\mathrm{prob}(D_1, D_2 | \mathrm{A}, I)}{\mathrm{prob}(D_1, D_2 | \mathrm{B}, I)} \approx \frac{(\mu_{\max} - \mu_{\min}) \, \sigma_{\max}}{\pi \sqrt{2}} \times \frac{N_1 \, N_2 \, (\sigma_o)^{2-N}}{N \, (\sigma_{o1})^{2-N_1} (\sigma_{o2})^{2-N_2}} . \qquad (4.31)$$

Since Bayes' theorem tells us that the ratio of the posterior probabilities for A and B is given by eqn (4.31) times $\mathrm{prob}(A | I)/\mathrm{prob}(B | I)$, which is usually taken as ~ 1, this formula provides a useful analytical solution to our classification problem; it should be fairly accurate as long as we have a reasonable amount of data ($N_1 \gg 1$ and $N_2 \gg 1$).

The preceding analysis can be generalized to the case when there are several sets of data $\{D_j\}$, with $j = 1, 2, \dots, M$. The evidence that all the measurements ($N = \sum N_j$) belong to the same classification-group is given by a double integral, like eqn (4.21), where the joint pdf $\mathrm{prob}(\{D_j\}, \mu, \sigma | A, I)$ is marginalized with respect to the unknown mean μ and standard deviation σ. The corresponding probability that each data-set relates to a different class takes the form of a product of the terms $\mathrm{prob}(D_j | B, I)$, which is just an extension of eqn (4.30):

$$\mathrm{prob}(\{D_j\} | \mathrm{B}, I) = \prod_{j=1}^{M} \mathrm{prob}(D_j | \mathrm{B}, I) .$$

Making the same simplifying approximations as before, eqn (4.31) becomes:

$$\frac{\mathrm{prob}(\{D_j\} | \mathrm{A}, I)}{\mathrm{prob}(\{D_j\} | \mathrm{B}, I)} \approx \left[\frac{(\mu_{\max} - \mu_{\min}) \, \sigma_{\max}}{\pi \sqrt{2}} \right]^{M-1} \times \frac{(\sigma_o)^{2-N}}{N} \times \prod_{j=1}^{M} \frac{N_j}{(\sigma_{oj})^{2-N_j}} ,$$

where σ_{oj} is the standard deviation, in eqn (4.25), of the N_j measurements D_j.

Returning to the case of just two data-sets, there is an important alternative question which we could ask. Suppose, for example, that we have to choose between two manufacturers of light bulbs on the basis of samples provided by them. Although we're interested in assessing the significance of the difference between their expected lifetimes (per unit cost), it's not really a classification problem. We already know that the products come from competing factories and, hence, will not have the same mean and standard deviation; the question is whether bulbs from one will last longer then those from the other. Thus, we need the posterior probability of the hypothesis $\mu_1 > \mu_2$; this can be evaluated from the joint pdf for the two means:

$$\text{prob}(\mu_1 > \mu_2 | D_1, D_2, I) = \int_0^\infty d\mu_1 \int_0^{\mu_1} d\mu_2 \, \text{prob}(\mu_1, \mu_2 | D_1, D_2, I) \,. \qquad (4.32)$$

If this probability was very close to unity, then we would buy from manufacturer 1; if it was almost zero, then the competitor would be better; a value of order 0.5 would indicate no strong preference for either.

Since a knowledge of the expected lifetime from one factory does not tell us much about the average performance of the other, the product rule of probability allows us to write the integrand in eqn (4.32) as

$$\text{prob}(\mu_1, \mu_2 | D_1, D_2, I) = \text{prob}(\mu_1 | D_1, I) \times \text{prob}(\mu_2 | D_2, I) \,. \qquad (4.33)$$

From our formulation of the problem in terms of means and standard deviations, the two pdfs on the right are just the marginal integrals of $\text{prob}(\mu_j, \sigma_j | D_j, I)$ with respect to σ_j; according to the analysis in Section 3.3, they yield the Student-t distribution of eqn (3.43). Substituting these into eqn (4.33), the double integral of eqn (4.32) can then be calculated numerically. If the number of samples provided is moderately large, the Student-t distributions will tend towards Gaussians, so that the joint pdf for the means can be approximated by

$$\text{prob}(\mu_1, \mu_2 | D_1, D_2, I) \approx \frac{\sqrt{N_1 N_2}}{2 \pi S_1 S_2} \exp\left(-\frac{1}{2}\left[\frac{N_1 (\mu_1 - \mu_{o1})^2}{S_1^2} + \frac{N_2 (\mu_2 - \mu_{o2})^2}{S_2^2} \right]\right) ,$$

where μ_{o1}, μ_{o2}, S_1 and S_2 are given by the appropriate summations over D_1 and D_2, respectively, in eqns (3.39) and (3.40). With a change of variables to $Z = \mu_1 - \mu_2$, as discussed in Section 3.6, the integral of eqn (4.32) reduces to

$$\text{prob}(\mu_1 > \mu_2 | D_1, D_2, I) \approx \frac{1}{S_z \sqrt{2\pi}} \int_0^\infty \exp\left[-\frac{(Z - z_o)^2}{2 S_z^2} \right] dZ \,, \qquad (4.34)$$

where $z_o = \mu_{o1} - \mu_{o2}$ and $S_z^2 = S_1^2/N_1 + S_2^2/N_2$. Thus, the integral properties of the Gaussian pdf tells us that the required probability will be about 0.84 if the difference between the two sample means z_o equals S_z, almost 0.98 for $z_o = 2 S_z$, and so on; it will be one minus these numbers if z_o is negative, but of the same magnitude.

It is interesting to note that eqn (4.34), unlike eqn (4.31), does not have an explicit dependence on the prior range $\mu_{\max} - \mu_{\min}$ and $\sigma_{\max}$. In fact, we've already seen this type of behaviour in Section 4.1. In the story of Mr A and Mr B, the posterior pdf for the parameter λ, $\mathrm{prob}(\lambda | D, B, I)$, is unaffected by the choice of $\lambda_{\min}$ and $\lambda_{\max}$ as long as they don't cause a significant truncation of the likelihood function; even if this is so, the prior range still plays an important rôle in the assessment of the relative merit of the two theories. The point is that in eqns (4.31) and (4.34) we are asking different questions of the data; we should not be surprised, therefore, if the answers are not the same.

Before moving on, let's consider how the analysis would have to be modified if the data $\{x_k\}$ were subject to uncertainties described by error-bars $\{\epsilon_k\}$. These are often neglected because they are small compared with the underlying variation defined by σ; for example, whereas the lifetime of a particular light bulb can be measured to a fraction of a second, it can differ from others within the sample by many hours. For the case of our intrepid archaeologist, however, the estimation of the brain size from skull fragments will be a less precise operation. This can be taken into account by expressing the likelihood function for the datum x_k as a marginal integral over the exact, but unknown, value of the sample $\hat{x}_k$:

$$\mathrm{prob}(x_k | \epsilon_k, \mu, \sigma, I) = \int \mathrm{prob}(x_k, \hat{x}_k | \epsilon_k, \mu, \sigma, I) \, \mathrm{d}\hat{x}_k$$

$$= \int \mathrm{prob}(x_k | \hat{x}_k, \epsilon_k, I) \, \mathrm{prob}(\hat{x}_k | \mu, \sigma, I) \, \mathrm{d}\hat{x}_k , \quad (4.35)$$

where unnecessary conditioning statements have been omitted in the second line. Assigning Gaussian pdfs for both the terms on the right-hand side, we obtain

$$\mathrm{prob}(x_k | \epsilon_k, \mu, \sigma, I) = \frac{1}{\sqrt{2\pi(\epsilon_k^2 + \sigma^2)}} \exp\left[-\frac{(x_k - \mu)^2}{2(\epsilon_k^2 + \sigma^2)} \right] . \quad (4.36)$$

The likelihood function for a whole set of independent measurements will then be a product of such terms; eqn (4.23) can be regarded as a special (but common) case, when $\epsilon_k \ll \sigma$. With this modification, the rest of the analysis proceeds as before. The problem can easily be tackled numerically, or even approximated analytically if all the error-bars are of the same size ($\epsilon_k = \epsilon$, for all k).

4.3.2 Luminescence dating

The next example comes from a dating technique, involving the *optically stimulated luminescence* from minerals, which is suitable for the study of sedimentary deposition and other geomorphological processes (Aitken 1998, Stokes 1999). For our purposes, we need say only that the data consist of N measurements $\{x_k\}$, of the 'equivalent dose' of laboratory radiation D_e, with error-bars $\{\epsilon_k\}$, such as those shown in Fig. 4.5(a). Our task is to infer the age structure of the sediment sample, $F(x)$. A subsequent conversion from x, or D_e, to an absolute chronology requires the additional collection of environmental radiation information (the annual dose rate).

This problem takes a model selection form if $F(x)$ can reasonably be approximated as the sum of a few, say M, discrete age components:

$$F(x) = \sum_{j=1}^{M} A_j f_j(x), \qquad (4.37)$$

where A_j is the relative contribution from the jth era of sediment deposition, $f_j(x)$, so that $\sum A_j = 1$. In the simplest case, which we adopt here, the f_j's will just be δ-functions: $f_j(x) = \delta(x - x_{oj})$. The prior for the amplitudes and equivalent doses (or locations) of the M components, $\text{prob}(\{A_j, x_{oj}\} \,|\, M, I)$, can be taken as being uniform in the range $0 \leqslant A_j \leqslant 1$, subject to normalization, and $0 \leqslant x_{oj} \leqslant x_{\max}$ (where $x_{\max}$ could be set by the largest datum).

As above, the likelihood for an individual measurement can be written as a marginal integral over its exact, but unknown, value $\hat{x}_k$:

$$\text{prob}(x_k \,|\, \epsilon_k, F(x), I) = \int \text{prob}(x_k \,|\, \hat{x}_k, \epsilon_k, I)\, F(\hat{x}_k)\, \mathrm{d}\hat{x}_k, \qquad (4.38)$$

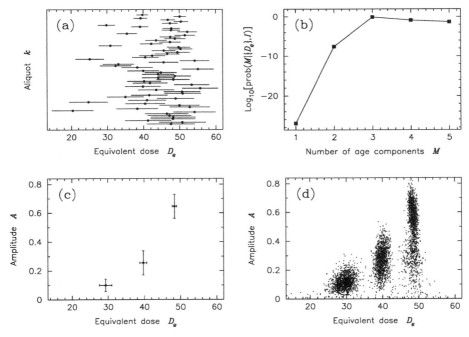

Fig. 4.5 (a) Luminescence dating measurements from 65 aliquots. (b) The logarithm of the posterior pdf for the number of age components. (c) The inferred proportions and equivalent doses of the optimal three age components, and their 1-σ error-bars. (d) Samples of proportions and equivalent doses drawn from their posterior pdf, generated by a Monte Carlo algorithm, where the number of age components has been marginalized out.

where we have used the definition of $F(x)$, $F(\hat{x}_k) = \mathrm{prob}(\hat{x}_k | F(x), I)$, on the right. With a Gaussian assignment for the noise, $\mathrm{prob}(x_k | \hat{x}_k, \epsilon_k, I)$, and a δ-function model for the f_j's in eqn (4.37), the likelihood function for x_k becomes

$$\mathrm{prob}(x_k | \epsilon_k, F(x), I) = \sum_{j=1}^{M} \frac{A_j}{\epsilon_k \sqrt{2\pi}} \exp\left[-\frac{(x_k - x_{oj})^2}{2\,\epsilon_k^2} \right]. \qquad (4.39)$$

The generalization to the case where each age component has a Gaussian width w_j is straightforward, and leads to a similar formula with the terms in the summation resembling those in eqn (4.36) with $\sigma = w_j$ (instead of zero). Making the usual assumption of independence, the joint likelihood of all the data is just a product of N such factors.

As noted in Section 4.1, the probabilistic evidence for a given model is determined principally by the mean value of the resulting likelihood function (averaged over the prior pdf). Its evaluation, within the quadratic approximations of Section 4.2, yields the posterior pdf for M shown in Fig. 4.5(b); the corresponding estimates of the proportions and equivalent doses of the optimal three age components are displayed in Fig. 4.5(c). If the same calculation was carried out with a Monte Carlo algorithm, such as that discussed in Chapter 9, with M marginalized out (using a Poisson prior with $\langle M \rangle = 1$), then we would obtain estimates of $\{A_j, x_{oj}\}$ indicated in Fig. 4.5(d). There is strong evidence, therefore, that the given sediment sample was a mixture of grains from three different ages. One possibility that could explain such post-depositional mixing is insect activity.

4.3.3 Interlude: what not to compute

The final example in this chapter comes from *quasi-elastic* neutron scattering, which is used to study the rotational and diffusive motions of atoms and molecules. The one-dimensional data, such as those shown in Fig. 4.6(a), are often analysed in terms of the sum of a few Lorentzian components, all centred at the origin; using the notation of eqn (4.12), the ideal underlying spectrum $G(x)$ takes the form

$$G(x) = A_O\,\delta(x) + \sum_{j=1}^{M} A_j\,\frac{w_j}{\pi\left(x^2 + w_j^2\right)}, \qquad (4.40)$$

where A_j and w_j are the amplitude and width of the jth line, and A_O is the δ-function contribution from elastic scattering. As in eqn (4.13), the measured neutron counts are a blurred and noisy version of this model and subject to a slowly varying background; the instrumental resolution function is shown in Fig. 4.6(b). Since each Lorentzian is associated with a particular type of molecular motion, it is important to ascertain how many components there is most evidence for in the data.

The problem of obtaining the best estimate of M is essentially the same as that addressed in Section 4.2; the main difference is that it is now the widths, rather than positions, of the excitations which are not known. By making very similar simplifying approximations, we are led to a formula just like eqn (4.20) except that $x_{\max} - x_{\min}$ is replaced with $w_{\max}$; a related algorithm is given in Sivia *et al.* (1992). Following this procedure yields the posterior pdf for M shown by the filled-in squares (and solid line) in Fig. 4.6(c); there is, therefore, most evidence for two quasi-elastic components. The best estimates of the amplitudes and widths of the Lorentzians, and their (1-σ) error-bars, are given in Fig. 4.6(d).

This illustration seems fairly straightforward and not unlike our earlier example. A *Fourier transform* of eqn (4.40), however, reveals that it is equivalent to the notorious problem of characterizing the sum of decaying exponentials. The difficulty is described graphically halfway through Acton's 1970 book, in a section entitled: 'Interlude: what *not* to compute'. He considers the analysis of radioactive decay. In the first instance, the abundance A and B of two known substances, with decay rates a and b, is to be estimated from the data

$$y(t) \;=\; A\,\mathrm{e}^{-at} + B\,\mathrm{e}^{-bt},$$

where y is proportional to the number of counts recorded at time t. Although this is easily dealt with through least-squares fitting, there is a closely-related situation which is far more troublesome than it looks. In the second case, the substances are not known;

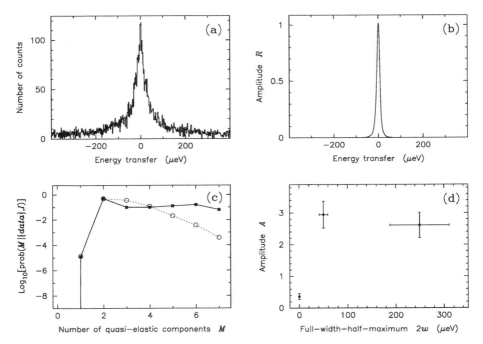

Fig. 4.6 (a) Data from a quasi-elastic neutron scattering experiment. (b) The instrumental resolution, or blurring, function. (c) The logarithm of the posterior pdf for the number of Lorentzian lines; the solid markers resulted from the use of a formula akin to eqn (4.20), whereas the open circles were obtained with a Monte Carlo algorithm. (d) The inferred amplitudes and widths of the quasi-elastic components, and their error-bars; $w=0$ corresponds to elastic scattering.

therefore, all four parameters (a, b, A and B) have to be inferred. Acton states that the solution to this problem lies in the chemical, rather than computer, laboratory because: '... it is well known that an exponential equation of this type in which all four parameters are to be fitted is *extremely* ill conditioned. That is, there are many combinations of (a, b, A, B) that will fit most exact data quite well indeed (will you believe four significant figures?) and when experimental noise is thrown into the pot, the entire operation becomes hopeless.' He·concludes: 'But those with Faith in Science do not always read the Book — and must be spanked or counselled.'

The sobering message of this anecdote is already contained in the solid line of Fig. 4.6(c): in contrast to the example of Fig. 4.4, the posterior pdf for the number of quasi-elastic components is much flatter after the maximum (at $M = 2$). Thus the analysis is automatically warning us that the most we can really say is that there are at least two Lorentzians. Indeed, even if we knew that there were more than two broadened lines, the (marginal) error-bars for their inferred parameters would be very large. The precise form of the solid line in Fig. 4.6(c) is itself suspect, because the simplifying approximations which underlie eqn (4.20) break down when the spread of the likelihood function becomes comparable to the prior range; a more reliable evaluation of the evidence integral is then given by a Monte Carlo algorithm, as illustrated by the dotted line in Fig. 4.6(c). Our conclusions are in general agreement with the folklore that it is difficult to characterize quasi-elastic scattering reliably if there are more than a couple of Lorentzian components. Once again, we see that Laplace's contention is borne out: probability theory is nothing but common sense reduced to calculation.

5 Assigning probabilities

In the preceding three chapters, we have seen how pdfs can be manipulated, with the sum and product rule of probability, to address data analysis problems; little was said, however, about their assignment in the first place. We now turn to this basic question. In addition to justifying the earlier use of the binomial, Gaussian and Poisson distributions, we will have our first encounter with the principle of maximum entropy.

5.1 Ignorance: indifference and transformation groups

In Section 1.2, we outlined how Cox showed that any method of plausible reasoning, which satisfies elementary requirements of logical consistency, must be equivalent to the use of probability theory. While the sum and product rule specify the relationship between pdfs, they do not tell us how to assign them. Are there any general principles to help us do this?

The oldest idea dates back to Bernoulli (1713), in what he called the 'principle of insufficient reason'; Keynes (1921) later renamed it the 'principle of indifference'. It states that if we can enumerate a set of basic, mutually exclusive, possibilities, and have no reason to believe that any one of these is more likely to be true than another, then we should assign the same probability to all. If we consider an ordinary die, for example, we can list the six potential outcomes of a roll,

$$X_i \equiv \text{the face on top has } i \text{ dots},$$

for $i = 1, 2, \ldots, 6$; according to Bernoulli, therefore, we have

$$\text{prob}(X_i|I) = 1/6,\tag{5.1}$$

where the background information I consists of nothing more than the enumeration of the possibilities. Although our everyday intuition tells us that this assignment is very reasonable, can we justify it in a more fundamental way? The reason for asking this question is that a better understanding of this easy problem might shed light on how to deal with more complicated situations later.

The most open-ended aspect of the statement of Bernoulli's principle above is what we mean by not having any reason to believe that one possibility is more likely to be true than another; it's worth trying to elaborate on this. Let's imagine that the potential outcomes of the die-roll were denoted by the first six letters of the alphabet; then, we need to assign the probabilities $\text{prob}(A|I)$, $\text{prob}(B|I)$, $\ldots$, $\text{prob}(F|I)$. This is, of course, nothing but a change of nomenclature: A could be equivalent to X_1, B to X_2, and so on. Having ascribed relative truth-values to propositions A to F, suppose that we were told that a mistake had been made and that A really stood for X_6, and B for X_5, etc.: Should such a reordering make any difference? If the conditioning on I represents a gross ignorance about the details of the situation, then the answer is no; the contrary would indicate that we were in possession of cogent information other than the simple enumeration of the possibilities. Consistency demands, therefore, that our probability assignment should not change if the order in which the propositions are listed is rearranged; the only way to satisfy this requirement is through eqn (5.1). This justification of Bernoulli's principle has led Jaynes (1978) to suggest that it is more apt to think of it as a consequence of the 'desideratum of consistency'; we will find that this type of argument forms the basis of the central theme in this chapter.

Bernoulli, himself, appreciated that the principle of insufficient reason could only be applied to a very limited number of problems; mainly, those involving games of chance. And, indeed, the scope of possible problems is far wider than those for which probability assignments are essentially definitive. Nevertheless, we can still obtain some important elementary results by combining the principle with the sum and product rules of probability. Consider, for example, the case of coloured balls being drawn randomly from an urn. If we knew that the contents consisted of W white balls and R red ones, and nothing more, then the principle of indifference tells us that we should assign the uniform pdf

$$\text{prob}(j|I) = \frac{1}{R+W} \, , \tag{5.2}$$

for the proposition that any particular ball, denoted by the index j, will be chosen. Marginalization and the product rule then allow us to express the probability that the colour drawn will be red as

$$\text{prob}(\text{red}|I) = \sum_{j=1}^{R+W} \text{prob}(\text{red}, j|I) = \frac{1}{R+W} \sum_{j=1}^{R+W} \text{prob}(\text{red}|j, I) \, ,$$

where we have substituted for $\text{prob}(j|I)$ from eqn (5.2) on the far right. The term $\text{prob}(\text{red}|j, I)$ will be equal to one if the jth ball is red and zero if it's white; the known contents of the urn, therefore, tell us that the summation will be equal to R. Hence, we obtain the anticipated result:

$$\text{prob}(\text{red}|I) = \frac{R}{R+W} \, . \tag{5.3}$$

We have derived eqn (5.3) from the assignment in eqn (5.2) with the sum and product rule, and it justifies the common notion of probability as

$$\text{prob}(\text{red}|I) = \frac{\text{number of cases favourable to red}}{\text{total number of equally possible cases}} \, .$$

Other familiar relationships emerge when we consider the result of repeating this ball-drawing procedure many times over. For algebraic simplicity, let us restrict ourselves to the case of 'sampling with replacement' where the contents of the urn are the same each time. What is then the probability that N such trials will result in r red balls? Using marginalization and the product rule, we can express this as

$$\text{prob}(r|N,I) = \sum_k \text{prob}(r, S_k|N, I)$$

$$= \sum_k \text{prob}(r|S_k, N, I)\,\text{prob}(S_k|N, I)\,, \qquad (5.4)$$

where the summation is over the 2^N possible sequences of red–white outcomes $\{S_k\}$ of N draws. The first term in the second line, $\text{prob}(r|S_k, N, I)$, will be equal to one if S_k contains exactly r red balls and zero otherwise; thus, we need only consider those sequences which have precisely r red outcomes for $\text{prob}(S_k|N, I)$. Since I assumes a general ignorance about the situation, other than a knowledge of the contents of the urn, the result of one draw does not influence what we can infer about the outcome of another; the probability of drawing any particular sequence S_k depends only on the total number of red (and complementary white) balls obtained, therefore, and not on their order. Specifically, for the $\{S_k\}$ which matter, we have

$$\text{prob}(S_k|N, I) = \left[\text{prob}(\text{red}|I)\right]^r \times \left[\text{prob}(\text{white}|I)\right]^{N-r}.$$

Substituting for $\text{prob}(\text{red}|I)$ from eqn (5.3), and for the corresponding probability of getting a white ball, this gives

$$\text{prob}(S_k|N, I) = \frac{R^r\,W^{N-r}}{(R+W)^N}\,. \qquad (5.5)$$

Hence, the summation of eqn (5.4) reduces to this term times the number of possible sequences of N-draws which contain exactly r red balls. To evaluate the latter, we must make a brief digression to the topic of *permutations* and *combinations*.

Let us begin this short foray into counting exercises with the easiest problem: In how many ways can n different objects be arranged in a straight line? Well, there are n choices for the first item, leaving $n-1$ possibilities for the second; this means that two objects can be picked in $n \times (n-1)$ ways, where their order matters. Continuing along this path, there are $n-2$ choices for the third item, $n-3$ for the fourth, and so on; the total number of permutations is then given by the product

$$n \times (n-1) \times (n-2) \times \cdots \times 3 \times 2 \times 1 = n!\,. \qquad (5.6)$$

A closely related question is as follows: In how many ways can we sequentially pick m objects from n different ones? We have, in fact, covered this case as an intermediate step in the calculation above; we just stop the product when m items have been chosen:

$$n \times (n-1) \times (n-2) \times \cdots \times (n-m+2) \times (n-m+1)\,.$$

This number of permutations is often written as ^nP_m and can be expressed as a ratio of two factorials by using eqn (5.6):

$$^{n}\mathrm{P}_{m} = \frac{n!}{(n-m)!} . \qquad (5.7)$$

The special case of $m = n$ does reduce to $n!$, as expected, because $0! = 1$; the latter follows from the substitution of $n = 1$ in $n! = (n-1)! \times n$. If we are not interested in the order in which the m objects are picked, then this can be taken into account by dividing eqn (5.7) by the number of ways that m items can be permuted; the resulting number of combinations is often denoted by $^{n}\mathrm{C}_{m}$:

$$^{n}\mathrm{C}_{m} = \frac{n!}{m!\,(n-m)!} . \qquad (5.8)$$

This formula also arises in the context of the *binomial* expansion for integer powers, where it appears as the coefficient in the relevant summation:

$$(a+b)^{N} = \sum_{j=0}^{N} \frac{N!}{j!\,(N-j)!}\, a^{j}\, b^{N-j} . \qquad (5.9)$$

It is useful to note that the sum on the right is equal to unity if $a + b = 1$; this result will be helpful in some of the following analysis. With $a = b = 1$, $\sum {}^{n}\mathrm{C}_{m} = 2^{n}$, where m goes from 0 to n, which represents the number of relevant possibilities.

Returning to the task of evaluating $\mathrm{prob}(r|N,I)$, the final ingredient we need is the number of different ways in which exactly r red balls can be drawn in N trials. To see how we can use the results derived above to calculate this, let's think of the problem in the following terms: imagine that the integers 1 to N have been written on separate small pieces of paper; if we select r of them, then the numbers chosen can be thought of as representing the draws on which a red ball was obtained. Thus, the sequences we require correspond to the number of ways of selecting r integers out of N where their order is irrelevant: hence, there are $^{N}\mathrm{C}_{r}$ of them. From our discussion of eqn (5.4), therefore, the results in eqns (5.5) and (5.8) combine to yield

$$\mathrm{prob}(r|N,I) = \frac{N!}{r!\,(N-r)!} \times \frac{R^{r}\,W^{N-r}}{(R+W)^{N}} . \qquad (5.10)$$

We can easily check that this pdf is normalized because the associated summation can be written in the form of eqn (5.9):

$$\sum_{r=0}^{N} \mathrm{prob}(r|N,I) = \sum_{r=0}^{N} \frac{N!}{r!\,(N-r)!}\, p^{r}\, q^{N-r} = (p+q)^{N},$$

where $p + q = 1$ since

$$p = \frac{R}{R+W} \quad \text{and} \quad q = \frac{W}{R+W} . \qquad (5.11)$$

The pdf of eqn (5.10) allows us to compute the frequency, r/N, with which we expect to observe red balls:

$$\left\langle \frac{r}{N} \right\rangle = \sum_{r=0}^{N} \frac{r}{N} \, \mathrm{prob}(r|N,I) = \sum_{r=1}^{N} \frac{(N-1)!}{(r-1)!\,(N-r)!} \, p^r \, q^{N-r},$$

where the lower limit of the sum has been changed to $r=1$ because the contribution of the $r=0$ term is zero. Taking a factor of p outside the summation, and letting $j=r-1$, we find that

$$\left\langle \frac{r}{N} \right\rangle = p \sum_{j=0}^{N-1} \frac{(N-1)!}{j!\,(N-1-j)!} \, p^j \, q^{N-1-j} = p\,(p+q)^{N-1}.$$

Substituting for p and q from eqn (5.11), we again obtain the anticipated result,

$$\left\langle \frac{r}{N} \right\rangle = \frac{R}{R+W}, \qquad (5.12)$$

where the expected frequency of red balls, in repetitions of the urn 'experiment', is equal to the probability of obtaining one in a single trial. A similar calculation for the variance of r/N shows that the mean-square deviation from eqn (5.12) is given by

$$\left\langle \left(\frac{r}{N} - p \right)^2 \right\rangle = \frac{p\,q}{N}. \qquad (5.13)$$

Since this becomes zero in the limit of infinite N, it verifies that Bernoulli's famous theorem of large numbers is obeyed:

$$\lim_{N \to \infty} \left(\frac{r}{N} \right) = \mathrm{prob}(\mathrm{red}|I). \qquad (5.14)$$

Although Bernoulli was able to derive this relationship for predicting the long-run frequency of occurrence from the probability assignment of eqn (5.3), his unfulfilled quest lay in the reverse process: What could one say about the probability of obtaining a red ball, in a single draw, given a finite number of observed outcomes? The answer to that question had to await Bayes and Laplace; as we have already seen in the first four chapters, it lies at the very heart of data analysis.

5.1.1 The binomial distribution

As in the case of the urn with red and white balls, or the coin-flipping example of Section 2.1, the outcome of some experiments can only take one of two values; for generality, we can call them 'success' and 'failure'. Even if there were also yellow, green and blue balls in the urn, we could still formulate the problem in such terms as long as our primary interest lay in just one colour; 'white' would then stand for 'not red'. Following the nomenclature of the preceding analysis, we can define:

$$\mathrm{prob}(\mathrm{success}|I) = p \qquad \text{and} \qquad \mathrm{prob}(\mathrm{failure}|I) = q = 1-p,$$

for the probability of success, and failure, in a single trial. The formulae of eqns (5.10) and (5.11), therefore, give the pdf of obtaining r successes in N trials as

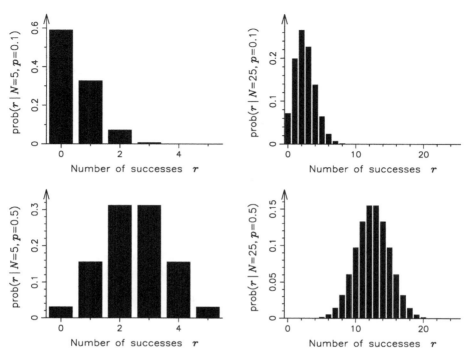

Fig. 5.1 The binomial distribution, $\text{prob}(r\,|\,N,p)$, or the probability of obtaining r favourable outcomes in N trials where p is the chance of success on any given attempt.

$$\text{prob}(r\,|\,N,I) \;=\; \frac{N!}{r!\,(N-r)!}\;p^r\,(1-p)^{N-r}\,, \tag{5.15}$$

where $r = 0, 1, 2, \ldots, N$. This is called the binomial distribution and is illustrated in Fig. 5.1 for two different values of p and N. The expected number of successes, $\langle r \rangle$, and the mean-square deviation from this average value, follows from eqns (5.11)–(5.13):

$$\langle r \rangle \;=\; Np \quad \text{and} \quad \left\langle (r - Np)^2 \right\rangle \;=\; Np\,(1-p)\,. \tag{5.16}$$

These results can be verified pictorially from the examples shown in Fig. 5.1.

5.1.2 Location and scale parameters

Having seen how some familiar elementary results can be derived from the principle of indifference, and sum and product rule of probability, let us return to the central topic of this chapter: namely, the assignment of probabilities. As stated earlier, Bernoulli's principle of insufficient reason can be used when we are able to enumerate a set of basic possibilities (and no more); this implicitly assumes that the quantity of interest, X, is restricted to certain discrete values, as in eqn (5.1). Can we generalize the argument to cover the case of continuous parameters?

Suppose that X represented the position of a lighthouse along a straight coast, as in Section 2.4. Then, given the information I, the probability that X lies in the infinitesimally small range between x and $x + \delta x$ is given by

$$\text{prob}(X = x \,|\, I)\,\mathrm{d}X = \lim_{\delta x \to 0} \text{prob}(x \leqslant X < x + \delta x \,|\, I)\,.$$

Although we treat continuous pdfs as the limiting case of discrete ones, the concept of enumerating the possibilities is rather awkward. Nevertheless, we can still make use the idea of consistency which underlies the principle of indifference. For example, if we were told that a mistake had been made in defining the origin, so that the position previously quoted as x was actually $x + x_0$, should this make any difference to the pdf assigned for X? If I indicates gross ignorance about the details of the situation, other than a knowledge that X pertains to a location (and some idea of its possible range), then the answer is 'not much'; the contrary would indicate that we were in possession of other cogent information regarding the position of the lighthouse. Consistency demands, therefore, that the pdf for X should change very little with the value of the offset x_0; this requirement can be written mathematically as

$$\text{prob}(X \,|\, I)\,\mathrm{d}X \approx \text{prob}(X + x_0 \,|\, I)\,\mathrm{d}(X + x_0)\,.$$

Since x_0 is a constant, $\mathrm{d}(X + x_0) = \mathrm{d}X$; this leads to the solution

$$\text{prob}(X \,|\, I) \approx \text{constant in the allowed range}\,, \tag{5.17}$$

and zero otherwise. Thus complete ignorance about a *location* parameter is represented by the assignment of a uniform pdf.

Another common problem concerns quantities which are associated with a size or magnitude, whereby it is the relative or fractional change that is important (rather than the absolute one, as in the case of a location parameter); these are often called *scale* parameters. For example, we could be interested in the length L of a biological molecule. What should we assign for the pdf $\text{prob}(L \,|\, I)$, where I represents gross prior ignorance about the value of L? Well, if we really had little idea about the length scale involved, then the graph of the pdf should be essentially invariant with respect to a stretching, or shrinking, of the horizontal L-axis. In other words, if we were told that a mistake had been made in the units of length quoted, so that they should have been ångstroms instead of nanometres, then this should not make much difference to the pdf we assign. This requirement of consistency can be written as

$$\text{prob}(L \,|\, I)\,\mathrm{d}L \approx \text{prob}(\beta L \,|\, I)\,\mathrm{d}(\beta L)\,,$$

where β is a positive constant; since $\mathrm{d}(\beta L) = \beta\,\mathrm{d}L$, it can only be satisfied by

$$\text{prob}(L \,|\, I) \propto 1/L \quad \text{in the allowed range}\,, \tag{5.18}$$

and zero otherwise. This pdf is often called a *Jeffreys' prior*, since it was first suggested by him (1939); it represents complete ignorance about the value of a scale parameter.

This result seems less weird once we realize that it's equivalent to a uniform pdf for the logarithm of L, $\mathrm{prob}(\log L \,|\, I) = \mathrm{constant}$, as can be verified with a change of variables according to eqn (3.77). In essence, it confirms our practical experience that problems involving magnitudes are best done using logarithmic graph paper!

5.2   Testable information: the principle of maximum entropy

We have now seen how some pdfs can be assigned when given only the nature of the quantities involved. The method hinges on the use of consistency arguments and requires a consideration of the *transformation groups* which characterize ignorance for the given situation. For the case of a finite set of discrete possibilities, this means that the related pdf has to be invariant with respect to any permutation of the relevant propositions. Continuous parameters are treated in the same way, with the appropriate transformations being an origin shift or an axis stretch for the two most common cases of location and scale variables. Having dealt with ignorance, let's move on to a more enlightened situation.

Suppose that a die, with the usual six faces, was rolled a very large number of times; if we were only told that the average result was 4.5, what probability should we assign for the various possible outcomes $\{X_i\}$ that the face on top had i dots? The information I provided can be written as a simple constraint:

$$\sum_{i=1}^{6} i \, \mathrm{prob}(X_i \,|\, I) = 4.5 \,. \tag{5.19}$$

Since the uniform pdf of eqn (5.1) predicts an average value of 3.5, instead of 4.5, it can be ruled out as a valid assignment. However, there are still many pdfs which are consistent with eqn (5.19); which one is best?

The constraint of eqn (5.19) is an example of *testable information*; with such a condition, we can either accept or reject (outright) any proposed pdf. Jaynes (1957) has suggested that, in this type of situation, we should make the assignment by using the principle of maximum entropy (MaxEnt): that is, we should choose that pdf which has the most entropy S while satisfying all the available constraints. Explicitly, for the case of the die above, we need to maximize

$$S = -\sum_{i=1}^{6} p_i \log_e[p_i] \,, \tag{5.20}$$

where $p_i = \mathrm{prob}(X_i \,|\, I)$, subject to normalization and the condition of eqn (5.19):

$$\sum_{i=1}^{6} p_i = 1 \qquad \text{and} \qquad \sum_{i=1}^{6} i \, p_i = 4.5 \,.$$

Such a constrained optimization can be done with the method of *Lagrange multipliers* (and some numerical analysis) and yields the pdf shown in Fig. 5.2.

At this juncture, it is reasonable to ask why the function in eqn (5.20) should be particularly favoured as a selection criterion. A good qualitative discussion of this question,

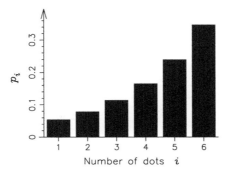

Fig. 5.2 The MaxEnt assignment for the pdf for the outcomes of a die roll, given only that it has the usual six faces and yields an average result of 4.5.

with reference to the die problem above, is given in Jaynes (1963); there he shows how the MaxEnt assignment reflects many of the properties we would expect from common sense. In a more general context, the entropy function can be justified in a variety of different ways: arguments ranging from information theory (Shannon 1948) to logical consistency (Shore and Johnson 1980), and several others besides, all lead to the conclusion that the '$-\sum p_i \log[p_i]$' criterion is highly desirable. To get a feel for why this is so, let us consider two simple examples. The first is the *kangaroo problem* of Gull and Skilling (1984), which they describe as a physicists' perversion of the formal mathematical analysis of Shore and Johnson; the second is a combinatorial argument, often phrased in terms of a hypothetical team of monkeys.

The kangaroo problem is as follows:

Information: A third of all kangaroos have blue eyes, and a quarter of all kangaroos are left-handed.

Question: On the basis of this information alone, what proportion of kangaroos are both blue-eyed and left-handed?

Well, for any given kangaroo, there are four distinct possibilities. Namely, that it is (1) blue-eyed and left-handed, (2) blue-eyed and right-handed, (3) not blue-eyed but left-handed or (4) not blue-eyed and right-handed. Following Bernoulli's law of large numbers, the expected values of the fraction of kangaroos with traits (1)–(4) will be equal to the probabilities we assign to each of these propositions. Denoting the latter by p_1, p_2, p_3 and p_4, respectively, we can represent the situation pictorially by using a 2×2 truth, or contingency, table shown in Fig. 5.3(a).

Although there are four possible combinations of eye-colour and handedness to be considered, the related probabilities are not independent of each other. In addition to the usual normalization requirement, $p_1 + p_2 + p_3 + p_4 = 1$, we also have two conditions on the marginal probabilities: $p_1 + p_2 = 1/3$ and $p_1 + p_3 = 1/4$. This means that the 2×2 contingency table of Fig. 5.3(a) can be parameterized by a single variable x, or p_1 (say), as is shown in Fig. 5.3(b). All such solutions, where $0 \leqslant x \leqslant 1/4$, satisfy the constraints of the testable information: Which one is 'best'?

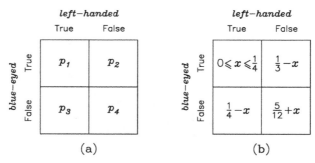

Fig. 5.3 The 2×2 truth table for the kangaroo problem: (a) the general case, where the probabilities for the four possibilities are denoted by p_1, p_2, p_3, p_4; (b) the corresponding parameterization in terms of a single variable x, which takes into account the information provided.

If a choice had to be made between the permissible pdfs, our common sense would draw us towards the assignment based on independence: $x = 1/12$. This is simply because any other value would indicate that a knowledge of the kangaroo's eye-colour told us something about its handedness; since we have no information to ascertain even the sign of any potential correlation, let alone its magnitude, this would seem to be somewhat foolhardy. There might well be a gene-linkage, of course, but we would be wise to shy away from assuming any strong connections ($x \to 0$ or $x \to 1/4$) without some pertinent evidence to support it.

The virtue of this little example is that it's easy enough to decide which is the most sensible pdf assignment, in the face of the inadequate information. This allows us to ask whether there is some function of the $\{p_i\}$ which, when maximized subject to the known constraints, yields the same preferred solution. If so, then it would be a good candidate for a general *variational principle* which could be used in situations that were too complicated for our common sense. To this end, Skilling (1988) has shown that the only functions which give uncorrelated assignments in general are those related monotonically to the entropy $S = -\sum p_i \log[p_i]$. The point is illustrated in Table 5.1, where the results obtained by using three proposed alternatives are listed; as expected, they all return optimal solutions with varying degrees of implied correlation between hand-

Table 5.1 The solutions to the kangaroo problem given by maximizing four different functions, subject to the constraints of the available information.

Variational function	Optimal x	Implied correlation
$-\sum p_i \log[p_i]$	0.0833	None
$-\sum p_i^2$	0.0417	Negative
$\sum \log[p_i]$	0.1060	Positive
$\sum \sqrt{p_i}$	0.0967	Positive

edness and eye-colour. This indicates that there is intrinsic merit in using the principle of maximum entropy for assigning pdfs; further support is provided by the so-called monkey argument.

5.2.1 The monkey argument

Suppose there are M distinct possibilities $\{X_i\}$ to be considered; our task is to ascribe truth-values to them, given some testable information I: $\mathrm{prob}(X_i|I) = p_i$. How can we do this in the most honest and fair way? Well, we could imagine playing a little game: the various propositions could be represented by different boxes, all of the same size, into which pennies are thrown at random; this job is often delegated to a hypothetical team of monkeys, to denote the fact that there should be no underlying bias in the procedure. After a very large number of coins have been distributed, the fraction found in each of the boxes gives a possible assignment for the probability for the corresponding $\{X_i\}$. The resulting pdf may not be consistent with the constraints of I, of course, in which case it should be rejected as a potential candidate; if it is in agreement, then it's a viable option. The boxes can then be emptied and the monkeys allowed to have another go at scattering the pennies. After many such trials, some distributions will be found to come up more often than others; the one that occurs most frequently (and satisfies I) would be a sensible choice for $\mathrm{prob}(\{X_i\}|I)$. This is because our ideal monkeys have no particular axe to grind, and so this favoured solution can be regarded as the one that best represents our state of knowledge: it agrees with all the testable information available while being as non-committal as possible towards everything else. Let's see how this corresponds to the pdf with the greatest value of '$-\sum p_i \log[p_i]$'.

After the monkeys have scattered all the pennies given to them, suppose that we find n_1 in the first box, n_2 in the second, and so on. The total number of coins N is then defined by the condition

$$N = \sum_{i=1}^{M} n_i, \qquad (5.21)$$

and we will assume that it is very large, and certainly much greater than the number of boxes: $N \gg M$. This distribution $\{n_i\}$ will give rise to a corresponding candidate pdf $\{p_i\}$ for the possibilities $\{X_i\}$:

$$p_i = n_i/N, \qquad (5.22)$$

where $i = 1, 2, \ldots, M$. Since every penny can land in any of the boxes, there are M^N number of different ways of scattering the coins amongst them; by design, each is equally likely to occur. All of these basic sequences are not distinct, however, as many yield the same distribution $\{n_i\}$. The expected frequency F with which a $\{p_i\}$ will arise, therefore, is given by

$$F(\{p_i\}) = \frac{\text{number of ways of obtaining } \{n_i\}}{M^N}. \qquad (5.23)$$

To evaluate the numerator, we can break up the calculation into a series of simpler steps. Let's start with box 1 and ask: In how many different ways can n_1 coins be chosen from a total of N? Following our earlier discussion of permutations and combinations, in Section 5.1, the answer is just the binomial coefficient $^NC_{n_1}$. Next, moving on to box 2, we can ask: In how many different ways can n_2 coins be selected from the remaining $N-n_1$? The answer is $^{N-n_1}C_{n_2}$. Continuing in this vein, we find that the numerator in eqn (5.23) is given by the product of M such binomial terms; using eqns (5.8) and (5.21), the required number reduces to

$$^NC_{n_1} \times {}^{N-n_1}C_{n_2} \times {}^{N-n_1-n_2}C_{n_3} \times \cdots \times {}^{n_M}C_{n_M} = \frac{N!}{n_1!\, n_2! \cdots n_M!}\,.$$

Substituting this expression for the numerator in eqn (5.23), we obtain

$$\log[F] = -N\log[M] + \log[N!] - \sum_{i=1}^{M} \log[n_i!]\,, \qquad (5.24)$$

where we have taken the logarithm of both sides to help deal with the excessively large numbers involved. In fact, the right-hand side can be simplified by using the *Stirling* approximation:

$$\log_e[n!] \approx n\log_e[n] - n\,,$$

which is valid as $n \to \infty$. As we approach this limit, eqn (5.24) reduces to

$$\log[F] = -N\log[M] + N\log[N] - \sum_{i=1}^{M} n_i \log[n_i]\,,$$

because two of the terms cancel out according to eqn (5.21). Finally, substituting for n_i from eqn (5.22), and using the fact that $\sum p_i = 1$, we are led to the result

$$\log[F] = -N\log[M] - N\sum_{i=1}^{M} p_i \log[p_i]\,. \qquad (5.25)$$

Since this is related monotonically to the expected frequency with which the monkeys will come up with the candidate pdf $\{p_i\}$, the assignment $\mathrm{prob}(\{X_i\}|I)$ which best represents our state of knowledge is the one that gives the greatest value for $\log[F]$ while being consistent with the testable information I. As M and N are constants, this is equivalent to the constrained maximization of the function S:

$$S = -\sum_{i=1}^{M} p_i \log[p_i]\,. \qquad (5.26)$$

Any monotonic function of S will, of course, lead to the same result. Therefore, neither the base of the logarithm, nor the precise value of N in eqn (5.25), is of any great concern and need not be specified.

The functional form of eqn (5.26) is the same as the quantity called 'entropy' in thermodynamics; for this reason, it too is referred to by that name. The similarity is far from being purely coincidental, however, as Jaynes has pointed out in many of his papers (e.g. 1983). He has shown how the familiar relationships of thermodynamics can easily be derived by treating the subject as an exercise in inferential calculus, with MaxEnt playing the central rôle of a variational principle for assigning pdfs subject to the macroscopic constraints like temperature, pressure and volume. This approach provides a fundamentally different perspective for several branches of physics.

5.2.2 The Lebesgue measure

In the monkey argument above, we took all the boxes to be of the same size; this seems reasonable, since we wanted to be fair towards all of the propositions that they represented. If we were dealing with an ordinary die, for example, then the principle of indifference would naturally lead us to assign equal prior weight to the six possible outcomes. Suppose that, for some perverse reason, the problem is posed in terms of just three hypotheses:

$$X_i \equiv \text{the face on top has} \begin{cases} i \text{ dots} & \text{for } i = 1, 2\,, \\ 3, 4, 5 \text{ or } 6 \text{ dots} & \text{for } i = 3\,. \end{cases}$$

Given (only) the six-sided nature of the die, we would then be inclined to make the box for X_3 four times as large as that for X_1 and X_2. How does this unevenness affect the preceding analysis?

Let's go back to the case of the M distinct possibilities, but adjust the size of the boxes so that the chances that a monkey will throw a penny into the ith one is m_i. We always have the condition that

$$\sum_{i=1}^{M} m_i = 1\,,$$

but the m_i are not necessarily equal; if they were, then the m_i would all be $1/M$. What is now the expected frequency F that the monkeys will throw n_1 coins in the first box, n_2 in the second, and so on? Well, it will be given by the number of different ways of scattering the N pennies which yield the distribution $\{n_i\}$ times the probability of obtaining such a sequence of throws:

$$F(\{p_i\}) = \frac{N!}{n_1!\, n_2! \cdots n_M!} \times m_1^{n_1}\, m_2^{n_2} \cdots m_M^{n_M}\,. \tag{5.27}$$

This is called a *multinomial* distribution, and is a generalization of its binomial ($M=2$) counterpart given in eqn (5.15). If all the m_i are equal, then the product of the M terms of the far right reduce to the reciprocal of M^N and, as required, we recover eqn (5.23). Taking the logarithm of eqn (5.27), and using Stirling's approximation, we find that eqn (5.25) now becomes

$$\log[F] = \sum_{i=1}^{M} n_i \log[m_i] - N \sum_{i=1}^{M} p_i \log[p_i]\,.$$

Substituting for n_i from eqn (5.22), we finally obtain

$$\frac{1}{N} \log[F] = -\sum_{i=1}^{M} p_i \log\left[\frac{p_i}{m_i}\right] = S. \tag{5.28}$$

In other words, we can regard the entropy formula of eqn (5.26) as a special case of this more general form (where $m_i = 1/M$). This is known by various names, including the Shannon–Jaynes entropy, the Kullback number and the *cross-entropy*; in the last context, it is often written with the opposite sign so that it has to be minimized!

Although the need for boxes of uneven size might seem rather contrived in our die example, Jaynes (1963) has pointed out that the generalization to the form of eqn (5.28) is necessary when we consider the limit of continuous parameters:

$$S = -\int p(x) \log\left[\frac{p(x)}{m(x)}\right] dx. \tag{5.29}$$

The Lebesgue *measure*, $m(x)$, ensures that the entropy expression is invariant under a change of variables, $x \rightarrow y = f(x)$, because both $p(x)$ and $m(x)$ transform in the same way. Essentially, the measure takes into account how the (uniform) bin-widths in x-space translate to a corresponding set of (variable) box-sizes in an alternative y-space.

To obtain a better understanding of the nature of $m(x)$, let's consider maximizing the entropy of eqn (5.29) subject to only the normalization condition $\int p(x)\,dx = 1$. This type of calculation is done most easily as the limit of its discrete counterpart; as such, the method of Lagrange multipliers tells us that we must maximize Q with respect to the $\{p_i\}$, treated as independent variables:

$$Q = -\sum_i p_i \log_e\left[\frac{p_i}{m_i}\right] + \lambda\left(1 - \sum_i p_i\right),$$

and subsequently determine λ from the condition $\sum p_i = 1$. Since the partial derivative $\partial p_i/\partial p_j = 0$ if $i \neq j$, we have

$$\frac{\partial Q}{\partial p_j} = -1 - \log_e\left[\frac{p_j}{m_j}\right] - \lambda = 0,$$

for all j; this leads to the trivial solution

$$p_j = m_j\,e^{-(1+\lambda)}.$$

The normalization requirement for $\{p_i\}$ fixes the Lagrange multiplier, and in the continuum limit

$$p(x) = \text{prob}(x\,|\,\text{normalization}) \propto m(x). \tag{5.30}$$

In other words, $m(x)$ is any multiple of the pdf which expresses complete ignorance about the value of x; thus, the transformation-group (invariance) arguments of Section 5.1 are appropriate for ascertaining the measure.

5.3   MaxEnt examples: some common pdfs

Having seen two of the ways by which we are led to MaxEnt, let us illustrate its use with a few examples. We have, in fact, just met the simplest situation where the testable information consists purely of the normalization condition. If the measure is uniform, then eqn (5.30) tells us that we should assign $\mathrm{prob}(x|I) = \mathrm{constant}$. For the case of M discrete possibilities $\{X_i\}$, this reduces to $\mathrm{prob}(X_i|I) = 1/M$, in accordance with Bernoulli's principle of insufficient reason.

The uniform pdf itself gives rise to others, when manipulated through the sum and product rules of probability. We saw this in Section 5.1, when considering the repeated drawing of red and white balls from an urn. Assuming the contents of the container, and our state of ignorance, to be the same on each trial, we were led to the binomial distribution; sampling without replacement, on the other hand, would yield the *hypergeometric* distribution. For the continuous case of the lighthouse problem in Section 2.4, we used the fact that the uniform pdf for the azimuth angle of the emitted beam translates to a Cauchy distribution for the position of the flashes detected on the coast; the relevant transformation was derived in Section 3.6. Now let's see how MaxEnt gives rise to some other commonly met pdfs.

5.3.1   Averages and exponentials

Suppose that the testable information, pertaining to some quantity x, consisted of a knowledge of the expectation value μ; this can be written as the constraint

$$\langle x \rangle = \int x \, \mathrm{prob}(x|I) \, \mathrm{d}x = \mu. \tag{5.31}$$

What should we now assign for $\mathrm{prob}(x|I)$? According to the principle of MaxEnt, we need that pdf which has the most entropy while satisfying the conditions of eqn (5.31) and normalization. As we said earlier, this optimization is best done as the limiting case of a discrete problem; explicitly, we need to find the maximum of Q:

$$Q = -\sum_i p_i \log_e\!\left[\frac{p_i}{m_i}\right] + \lambda_0\!\left(1 - \sum_i p_i\right) + \lambda_1\!\left(\mu - \sum_i x_i \, p_i\right),$$

where λ_0 and λ_1 are Lagrange multipliers. Setting $\partial Q/\partial p_j = 0$, we obtain

$$p_j = m_j \, \mathrm{e}^{-(1+\lambda_0)} \, \mathrm{e}^{-\lambda_1 x_j}. \tag{5.32}$$

In fact, with $m_i = 1/6$ and $x_i = i$, this is the solution to the die problem plotted in Fig. 5.2; λ_0 and λ_1 were calculated numerically, so that the resultant pdf satisfied the requirements $\sum i\, p_i = 4.5$ and $\sum p_i = 1$. Generalizing to the continuous case, with a uniform measure, eqn (5.32) becomes a simple exponential function:

$$\mathrm{prob}(x|I) \propto \exp\!\left[-\lambda_1 x\right].$$

The normalization constant and λ_1 can easily be evaluated if the limits of integration, in eqn (5.31), are 0 and ∞; it results in the assignment

$$\text{prob}(x|\mu) = \frac{1}{\mu} \exp\left[-\frac{x}{\mu}\right] \quad \text{for } x \geqslant 0. \tag{5.33}$$

A uniform measure is not always the most appropriate, and this gives rise to other important pdfs; we'll illustrate this with the binomial and Poisson distributions shortly. Another example is given in Sivia and David (1994), where they show how both the *Wilson* distributions of crystallography emerge from the same constraint on the mean; it simply requires a careful consideration of the relevant measures. Now let's see what happens as we obtain some more cogent testable information.

5.3.2 Variance and the Gaussian distribution

Suppose that we know not only a value μ, but also the variance σ^2 about that value:

$$\left\langle (x-\mu)^2 \right\rangle = \int (x-\mu)^2 \,\text{prob}(x|I)\,\mathrm{d}x = \sigma^2. \tag{5.34}$$

To assign $\text{prob}(x|I)$, we need to maximize its entropy subject to normalization and eqn (5.34). For the discrete case, this is equivalent to finding the extremum of Q:

$$Q = -\sum_i p_i \log_e\left[\frac{p_i}{m_i}\right] + \lambda_0\left(1 - \sum_i p_i\right) + \lambda_1\left(\sigma^2 - \sum_i (x_i-\mu)^2 p_i\right),$$

where λ_0 and λ_1 are Lagrange multipliers. Setting $\partial Q/\partial p_j = 0$, we obtain

$$p_j = m_j\,\mathrm{e}^{-(1+\lambda_0)}\,\mathrm{e}^{-\lambda_1(x_j-\mu)^2}.$$

With a uniform measure, this generalizes into the continuum assignment

$$\text{prob}(x|I) \propto \exp\left[-\lambda_1(x-\mu)^2\right].$$

The normalization constant and λ_1 are readily evaluated if the limits of integration, in eqn (5.34), are $\pm\infty$; it results in the standard Gaussian pdf

$$\text{prob}(x|\mu,\sigma) = \frac{1}{\sigma\sqrt{2\pi}} \exp\left[-\frac{(x-\mu)^2}{2\sigma^2}\right], \tag{5.35}$$

where σ is defined in eqn (5.34), and μ is seen to be the mean as defined in eqn (5.31). This indicates why there is a close link between the normal distribution and quantities characterized solely by their mean and variance: the former is the most honest description of our state of knowledge, when given nothing more than this information about the latter.

So far, we have only considered pdfs of a single variable; the MaxEnt analysis is easily extended to the case of several parameters by expressing the entropy of eqn (5.29) as a multi-dimensional integral:

$$S = -\iint\cdots\int p(\boldsymbol{x}) \log\left[\frac{p(\boldsymbol{x})}{m(\boldsymbol{x})}\right] \mathrm{d}^N\boldsymbol{x}, \tag{5.36}$$

where $p(\boldsymbol{x}) = \mathrm{prob}(x_1, x_2, \ldots, x_N \,|\, I)$, and so on. If the testable information pertaining to the quantities $\{x_k\}$ consists of a knowledge of just their individual variances,

$$\left\langle (x_k - \mu_k)^2 \right\rangle = \iint \cdots \int (x_k - \mu_k)^2 \, p(\boldsymbol{x}) \, \mathrm{d}^N \boldsymbol{x} = \sigma_k^2 \,, \qquad (5.37)$$

where $k = 1, 2, \ldots, N$, then it can be shown that the maximization of eqn (5.36), with a uniform measure, yields a simple product of Gaussian pdfs:

$$\mathrm{prob}(\{x_k\} \,|\, \{\mu_k, \sigma_k\}) = \prod_{k=1}^{N} \frac{1}{\sigma_k \sqrt{2\pi}} \, \exp\left[-\frac{(x_k - \mu_k)^2}{2\,\sigma_k^2} \right]. \qquad (5.38)$$

This is, in fact, the same pdf as a least-squares likelihood function! If we identify the $\{x_k\}$ as the data $\{D_k\}$, with error-bars $\{\sigma_k\}$, and the $\{\mu_k\}$ as the predictions $\{F_k\}$ based on some given model, then eqn (5.38) corresponds to eqns (3.65) and (3.66).

From the MaxEnt point of view, the least-squares likelihood does not imply the existence (or assumption) of a mechanism for generating independent, additive, Gaussian noise. It is just seen as the pdf which best represents our state of knowledge given only the value of the expected square-deviation between our predictions and the data, as in eqn (5.37). This potential mismatch could stem from uncertainties in the measurement process, perhaps due to a poor calibration, but it may simply reflect the inadequacies of the simplifying approximations in the modelling of the situation. If we had cogent information about the covariance $\left\langle (x_i - \mu_i)(x_j - \mu_j) \right\rangle$, where $i \neq j$, then MaxEnt would assign a correlated multivariate Gaussian pdf for $\mathrm{prob}(\{x_k\} \,|\, I)$.

In Section 3.5, we mentioned the possibility of using the l_1-norm of eqn (3.72) as a criterion for fitting functional models to data. This involves the minimization of the sum of the moduli of the misfit residuals, rather than their squares; the latter is, of course, just χ^2 (or the l_2-norm) of eqn (3.66). This procedure follows naturally from MaxEnt if the testable information consists only of the expected value of the modulus of the discrepancy between theory and experiment for the individual data:

$$\left\langle \,|x_k - \mu_k|\, \right\rangle = \iint \cdots \int |x_k - \mu_k| \, p(\boldsymbol{x}) \, \mathrm{d}^N \boldsymbol{x} = \epsilon_k \,, \qquad (5.39)$$

where $k = 1, 2, \ldots, N$. Then, it can be shown that the maximization of eqn (5.36), with a uniform measure, yields the product of symmetric exponential pdfs:

$$\mathrm{prob}(\{x_k\} \,|\, \{\mu_k, \epsilon_k\}) = \prod_{k=1}^{N} \frac{1}{2\epsilon_k} \, \exp\left(-\frac{|x_k - \mu_k|}{\epsilon_k} \right). \qquad (5.40)$$

The logarithm of the likelihood function is, therefore, given by the l_1-norm (plus an additive constant). Having seen how the precise nature of the testable constraints can influence the pdf we assign, let's look at a couple of examples where the measure plays an equally important rôle.

5.3.3   MaxEnt and the binomial distribution

Although we have already derived the binomial distribution from elementary considerations in Section 5.1, it's instructive to see how it emerges from the use of the MaxEnt
principle. Suppose that we are given (only) the expected number of success in M trials,
$\langle N \rangle = \mu$. What should we assign for the probability of a specific number of favourable
outcomes, $\text{prob}(N|M,\mu)$?

According to the MaxEnt principle, we need to maximize the entropy S of eqn
(5.28) subject to the testable information

$$\langle N \rangle \;=\; \sum_{N=0}^{M} N \, \text{prob}(N|M,\mu) \;=\; \mu \tag{5.41}$$

and normalization. Following an earlier calculation, this optimization yields the pdf of
eqn (5.32); specifically, we obtain

$$\text{prob}(N|M,\mu) \;\propto\; m(N)\, e^{-\lambda N}, \tag{5.42}$$

where λ is a Lagrange multiplier and $m(N)$ is the measure. The former is, of course,
determined by the constraint on the mean, but we must first assign $m(N)$.

From our discussion of the Lebesgue measure in Section 5.2.2, $m(N)$ is proportional to the pdf which reflects gross ignorance about the details of the situation. Given
only that there are M trials, the principle of indifference tells us to assign equal probability to each of the 2^M possible outcomes. The number of different ways of obtaining
N successes in M trials, or $^{M}C_{N}$ in eqn (5.8), is therefore an appropriate measure for
this problem:

$$m(N) \;=\; \frac{M!}{N!\,(M-N)!}\,. \tag{5.43}$$

All we need to do now is to substitute this into eqn (5.42), and to impose the constraints
of normalization and eqn (5.41). The related algebra is simplified by noting that

$$\sum_{N=0}^{M} m(N)\, e^{-\lambda N} \;=\; \left(e^{-\lambda}+1\right)^{M}, \tag{5.44}$$

which follows from eqn (5.9) on putting $a = e^{-\lambda}$ and $b = 1$. Its reciprocal yields the
constant of proportionality in eqn (5.42), and its implicit differentiation with respect to
λ, giving

$$\sum_{N=0}^{M} N\, m(N)\, e^{-\lambda N} \;=\; M\left(e^{-\lambda}+1\right)^{M-1} e^{-\lambda},$$

allows eqn (5.41) to be reduced to $M\left(1+e^{\lambda}\right)^{-1} = \mu$. A little algebraic rearrangement
then shows eqn (5.42) to be a binomial pdf:

$$\text{prob}(N|M,\mu) \;=\; \frac{M!}{N!\,(M-N)!} \left(\frac{\mu}{M}\right)^{N} \left(1 - \frac{\mu}{M}\right)^{M-N}. \tag{5.45}$$

5.3.4 Counting and Poisson statistics

Many problems in science involve the counting of discrete events in a finite interval. These can include a temporal aspect, such as the number of neutrons (or X-ray photons, or buses, or accidents) detected in a given amount of time, or a spatial component, like the number of quasars in a certain patch of the sky. What should we assign for the probability of observing N events, given only the expected value $\langle N \rangle = \mu$?

This case is actually very similar to the preceding one, because the situation can be regarded as the $M \to \infty$ limit of Section 5.3.3. That is to say, we can imagine the finite interval of interest to be made up of a huge number of microscopic sub-intervals within each of which only a single event could occur. We will show how the binomial pdf of eqn (5.45) tends to a Poisson pdf as $M \to \infty$ in the next section, but let's see how the result can be obtained directly through MaxEnt by using the large-M limit of the measure in eqn (5.43):

$$m(N) = \frac{M^N}{N!} . \qquad (5.46)$$

The substitution of this measure into eqn (5.42) gives

$$\text{prob}(N|\mu) = A \frac{\left(M e^{-\lambda}\right)^N}{N!} , \qquad (5.47)$$

where A is a normalization constant; the imposition of this constraint gives

$$\sum_{N=0}^{\infty} \text{prob}(N|\mu) = A e^{M e^{-\lambda}} = 1 ,$$

where the summation has been carried out by noting that it is equivalent to the Taylor series expansion of $\exp(X)$, for $X = M e^{-\lambda}$. This allows eqn (5.47) to be written as

$$\text{prob}(N|\mu) = \frac{1}{e^{M e^{-\lambda}}} \frac{\left(M e^{-\lambda}\right)^N}{N!} . \qquad (5.48)$$

We must also satisfy the requirement that $\langle N \rangle = \mu$, and this leads to the condition $M e^{-\lambda} = \mu$. Substituting this into eqn (5.48), we obtain the Poisson distribution

$$\text{prob}(N|\mu) = \frac{\mu^N e^{-\mu}}{N!} , \qquad (5.49)$$

which is independent of the exact degree of subdivision of the original interval; in using the measure of eqn (5.46), we have implicitly assumed that it is large. The pdf of eqn (5.49) was illustrated in Fig. 3.2, for two different values of μ.

5.4 Approximations: interconnections and simplifications

The Poisson distribution can also be derived as the limiting form of a binomial pdf, where the probability of success in any given trial is very small but the number of attempts is extremely large. The probability of obtaining N successes in M trials, where the expected number of favourable outcomes $\langle N \rangle = \mu$, is given in eqn (5.45). As M becomes very large, $M!/(M-N)!$ can be approximated by M^N and the $M-N$ in the exponent of $(1-\mu/M)$ can be replaced by just M:

$$\mathrm{prob}(N\,|\,M,\mu) \;\approx\; \frac{M^N}{N!} \left(\frac{\mu}{M}\right)^N \left(1 - \frac{\mu}{M}\right)^M .$$

After cancelling the factors of M^N, the term on the far right can be recognized as one which tends to $\mathrm{e}^{-\mu}$ as $M \to \infty$; this is easily verified by considering its logarithm:

$$\log_e\left[\left(1 - \frac{\mu}{M}\right)^M\right] \;=\; M \log_e\left(1 - \frac{\mu}{M}\right) \;=\; -\mu - \frac{\mu^2}{2M} - \frac{\mu^3}{3M^2} - \cdots ,$$

where we have used the Taylor series expansion of $\log_e(1-X)$, for $X = \mu/M$, in the last step. Thus, we obtain the Poisson distribution of eqn (5.49).

This alternative derivation is useful because it tells us that the binomial pdf can be approximated by the Poisson formula

$$\mathrm{prob}(r\,|\,N,p) \;=\; \frac{N!}{r!\,(N-r)!}\; p^r\,(1-p)^{N-r} \;\approx\; \frac{(Np)^r\,\mathrm{e}^{-Np}}{r!} , \qquad (5.50)$$

if p is small but N is large. As a simple example, consider the case of a fruit grower who finds that 1 in 50 apples picked are bad: If 100 are sent out without sorting, what is the probability that r will be rotten? Since an individual apple can either be good or bad, this situation can be treated as a binomial problem with $N = 100$ and $p = 0.02$; from eqn (5.50), it can also be approximated by a Poisson pdf with $\mu = Np = 2$. The corresponding results given by the two pdfs, for $r = 0, 1, 2, \ldots, 7$, are shown in Table 5.2; as expected, there is good agreement between the two formulae.

Pursuing the theme of simplifications a little further, we find that the Poisson pdf can be approximated by a Gaussian expression if the average value μ is large. To see this, consider the logarithm of eqn (5.49):

Table 5.2 The Poisson approximation to the binomial pdf: the probability of finding r rotten apples in a carton of 100, given that 1 in 50 tend to be bad.

Number of rotten apples r	Binomial probability $\mathrm{prob}(r\,|\,N=100, p=0.02)$	Poisson approximation $\mathrm{prob}(r\,|\,\mu=2)$
0	0.1326	0.1353
1	0.2707	0.2707
2	0.2734	0.2707
3	0.1823	0.1804
4	0.0902	0.0902
5	0.0353	0.0361
6	0.0114	0.0120
7	0.0031	0.0034

$$L \;=\; \log_{\mathrm{e}}\big[\operatorname{prob}(N|\mu)\big] \;=\; N \log_{\mathrm{e}}(\mu) - \mu - \log_{\mathrm{e}}(N!) \,. \qquad (5.51)$$

Since the most probable values of N will be in the vicinity of μ, it is appropriate to use Stirling's formula if $\mu \gg 1$:

$$\log_{\mathrm{e}}(N!) \;\approx\; N \log_{\mathrm{e}}(N) - N + \tfrac{1}{2} \log_{\mathrm{e}}(2\pi N) \,.$$

The last term on the right can be ignored in the limit $N \to \infty$, as we did earlier, but its inclusion ensures a good approximation even for fairly small N; substituting this for $\log_{\mathrm{e}}(N!)$ in eqn (5.51), we obtain

$$L \;\approx\; N - \mu - N \log_{\mathrm{e}}(N/\mu) - \tfrac{1}{2} \log_{\mathrm{e}}(2\pi N) \,. \qquad (5.52)$$

To investigate the behaviour of L in the neighbourhood of μ, let's put $N = \mu + \epsilon$ and consider the case $|\epsilon| \ll \mu$. After a little rearrangement, L can be written as

$$L \;\approx\; -\tfrac{1}{2} \log_{\mathrm{e}}(2\pi\mu) + \epsilon - \big(\mu + \epsilon + \tfrac{1}{2}\big) \log_{\mathrm{e}}\!\left[1 + \frac{\epsilon}{\mu}\right] \,.$$

A Taylor series expansion of $\log_{\mathrm{e}}(1 + \epsilon/\mu)$ then yields

$$L \;\approx\; -\tfrac{1}{2} \log_{\mathrm{e}}(2\pi\mu) - \frac{\epsilon^2}{2\mu} - \frac{\epsilon}{2\mu} + \cdots \,,$$

where only the most significant terms have been retained. The variation of L in the vicinity of $N \approx \mu$ can ascertained by completing the square for ϵ; its exponential yields the desired Gaussian expression:

$$\operatorname{prob}(N|\mu) \;=\; \frac{\mu^{N} e^{-\mu}}{N!} \;\approx\; \frac{1}{\sqrt{2\pi\mu}} \, \exp\!\left[-\frac{\big(N - \mu + \tfrac{1}{2}\big)^2}{2\mu}\right] \,, \qquad (5.53)$$

for large μ. This tendency towards a normal distribution is evident in Fig. 3.2, even for $\mu = 12.5$; the correspondence improves as the average value gets bigger. If we had carried out the analysis above by substituting for μ instead of N in eqn (5.52), then we would have been led to an alternative formula:

$$\frac{\mu^{N} e^{-\mu}}{N!} \;\approx\; \frac{1}{\sqrt{2\pi N}} \, \exp\!\left[-\frac{(\mu - N)^2}{2N}\right] \,, \qquad (5.54)$$

which too is valid in the limit of a large number of counts ($N \approx \mu \gg 1$).

In the Gaussian approximation to the Poisson pdf of eqn (5.53), it can be seen that the variance of N is equal to μ; this is, in fact, a general property of the Poisson distribution. That is to say, it can be shown that $\operatorname{prob}(N|\mu)$ in eqn (5.49) satisfies

$$\big\langle (N - \mu)^2 \big\rangle \;=\; \sum_{N=0}^{\infty} (N - \mu)^2 \, \operatorname{prob}(N|\mu) \;=\; \mu \,, \qquad (5.55)$$

irrespective of the magnitude of μ. A calculation similar to that leading to eqn (5.53), but starting from eqn (5.15), yields the result

$$\text{prob}(r|N) = \frac{N!}{r!\,(N-r)!}\,p^r\,(1-p)^{N-r} \approx \frac{1}{\sigma\sqrt{2\pi}}\,\exp\!\left[-\frac{(r-\mu)^2}{2\,\sigma^2}\right], \quad (5.56)$$

where $\mu = Np \gg 1$ and $\sigma^2 = Np\,(1-p) \gg 1$. In other words, a binomial pdf can also be approximated by a Gaussian if both the mean and variance in eqn (5.16) are large. This was first shown by de Moivre (1733), for the case $p = 1/2$, and later by Laplace in greater generality; it can be verified pictorially from Fig. 5.1. The validity of eqns (5.53), (5.54) and (5.55) is often the justification for the use of a least-squares likelihood function, even when the most appropriate pdf is not intrinsically Gaussian.

5.5 Hangups: priors versus likelihoods

In Section 1.4, we noted that Laplace's approach to probability was rejected by many, soon after his death, on the grounds that a degree of belief was too vague a concept to be the basis of a rigorous theory. Even today, a discussion of Bayesian methods is usually qualified by the words 'subjective probabilities'. Like Jaynes, we believe that this notion is misguided. To explain this, let us try to get to the heart of the matter.

The main point of concern tends to be the choice of the prior pdf: What should we do if this is not known? This issue is often regarded as a major obstacle for the general use of the Bayesian approach and is cited as proof of its subjective nature. From our point of view, however, the question is itself rather strange. No probability, whether prior, likelihood or whatever, is ever 'known'; it is simply an assignment which reflects the relevant information that is available. Thus $\text{prob}(x|I_1) \neq \text{prob}(x|I_2)$, in general, where the conditioning statements I_1 and I_2 are different. Nevertheless, objectivity can, and ought to, be introduced by demanding that two people with the same information I should assign the same pdf; this requirement of consistency has been the central theme of this chapter. We have seen how invariance arguments, under the appropriate transformation groups, can often determine a pdf uniquely, when given only the nature of the quantities concerned; MaxEnt provides a powerful extension, and allows us to deal with the case of testable constraints. The point is that nowhere in Sections 5.1–5.4 have we explicitly differentiated between a prior and a likelihood; we have merely considered how to assign $\text{prob}(X|I)$ for different types of I. If X pertains to data, then we can call $\text{prob}(X|I)$ a likelihood; if neither X nor I refer to (new) measurements, then we may say it's a prior. The distinction between the two cases is one of nomenclature, and not that of objectivity or subjectivity. If it appears otherwise, then this is because we are usually prepared to state the conditioning assumptions for the likelihood function but shy away from doing likewise for the prior pdf.

5.5.1 Improper pdfs

At a technical level, the issue of *improper* priors is often quoted as a serious impediment for the general use of the Bayesian approach; that is to say, some pdfs which represent complete ignorance cannot be normalized. The Jeffreys' prior is most frequently cited as an example of the difficulty: given that a scale parameter can lie in the range 0 to ∞, there is no proportionality constant which will make the integral of the pdf in eqn (5.18) equal to unity. How can we get around this problem?

When faced with such a dilemma in an ordinary mathematical setting, we tend to excise the source of the difficulty and consider how the result from the remaining part evolves as the excluded quantity is gradually readmitted. We can do the same thing here: carry out the calculation for a proper prior, with a lower and upper bound of A and B, and investigate how the posterior pdf is affected as these limits approach the improper values. For the case of a Jeffreys' prior, for example, we would use

$$
\text{prob}(L\,|\,I) \;=\;
\begin{cases}
\dfrac{1}{\log_e(B/A)} \times \dfrac{1}{L} & \text{for } A \leqslant L < B\,, \\[2mm]
0 & \text{otherwise}\,,
\end{cases}
$$

where $A > 0$ and B is finite, multiply it by the likelihood $\text{prob}(data\,|\,L,I)$, normalize the resulting posterior pdf, and consider what happens as $A \to 0$ and $B \to \infty$. While the prior itself might diverge with these limits, the posterior $\text{prob}(L\,|\,data,I)$ is usually well-behaved. If the conclusions drawn from the posterior do depend strongly on the values of A and B, then probability theory is warning us that there is very little relevant information in the data; it is then not surprising that our prior knowledge, which must define these limits, should play a crucial rôle in the inference procedure.

After going through the formalities for handling improper priors, we often find that the same result would have been obtained if the unnormalized pdf (e.g. $1/L$) had been substituted directly into Bayes' theorem; while not wishing to encourage such complacency, the temptation for taking this short cut is understandable. In fact, we do this every time the maximum likelihood method, or the least-squares procedure, is used: both rely on the implicit use of a uniform prior, which cannot be normalized without defining a suitable range. In general, for parameter estimation, improper priors merely constitute a technical inconvenience rather than a serious difficulty. For model selection, however, the situation is quite different. As we saw in Chapter 4, the prior pdf must be normalized properly in that case (by setting suitable bounds if necessary) because the evaluation of the probabilistic evidence entails an averaging of the likelihood function over it.

5.5.2 Conjugate and reference priors

In order to address the concerns voiced over the subjectivity of the prior pdf, several procedures have been proposed for their 'automatic' assignment based on the nature of the likelihood function. For example, *conjugate* priors, which ensure that the functional form of the resultant posterior pdf is the same as that of the prior (e.g. see Bernardo and Smith 1994), have been advocated. While these are motivated by the pragmatic desire to have simplified algebra, an information-based criterion, leading to *reference* priors, has been put forward by Bernardo (1979). The idea is that, 'even for small sample sizes, the *information provided by the data should dominate the prior information*, because of the "vague" nature of the prior knowledge.'

Although the development of conjugate and reference priors can seem attractive, because they provide 'off the shelf' pdfs, we have some basic misgivings about them.

The principal concern is that they perpetuate the myth that the likelihood function, unlike the prior pdf, is inherently objective. From our point of view, this is not so: both are simply assignments that reflect states of knowledge. While we might expect our initial understanding of the object of interest to have a bearing on the experiment we conduct, it seems strange that the choice of the prior pdf should have to wait for, and depend in detail upon, the likelihood function.

Part II

Advanced topics

'Data analysis is simply a dialogue with the data.'
— Stephen F. Gull, Cambridge 1994

6 Non-parametric estimation

In Chapters 2 and 3, we dealt with problems which entailed the estimation of the values of a few well-defined parameters; the analysis was then extended, in Chapter 4, to cover the case where there was some doubt as to their optimal number. Now let's move on to the most difficult situation, in which we want to draw inferences about an object of interest but don't have enough information, or confidence, to be able to characterize it by a specific type of functional model. This is a rather open-ended question that requires us to think of both a suitable description and any weak constraints which might be appropriate for a flexible 'free-form' solution.

6.1 Introduction: free-form solutions

Many data analysis problems in science can be stated in the following general terms: Given a set of data $\{D_k\}$, and the relevant background information I, what can we say about the object of interest $f(x)$? Here the variable x is used rather loosely and could refer to a coordinate that was discrete, continuous or even multidimensional. For example, an astronomer might want to know the radio-brightness distribution of the two-dimensional sky, having measured the (complex) *visibilities* with an interferometer; a materials scientist could be faced with the task of estimating the three-dimensional preferred-orientation distribution of crystallites in a processed metal, given the number of neutrons that a sample scatters through various angles. As we have noted from the outset, our inference is summarized by the posterior pdf $\mathrm{prob}[f(x)|\{D_k\}, I]$; before we can even think about assigning the related prior and likelihood, however, we must first decide on a suitable way of describing $f(x)$. There are essentially three possibilities; let us consider each in turn.

If we know a lot about the object of interest, then we can often characterize it by a few specific parameters. Taking $f(x)$ to be a one-dimensional spectrum, for the sake of argument, a simple example would be:

$$f(x) = a_1\,\delta(x-x_1) + a_2\,\delta(x-x_2)\,, \tag{6.1}$$

where $\delta(x-x^*)$ is an infinitely sharp spike, of unit area, at $x = x^*$. Our inference about $f(x)$ is, therefore, embodied in the posterior pdf for the positions and amplitudes of the two δ-functions in eqn (6.1): $\mathrm{prob}(a_1, a_2, x_1, x_2|\{D_k\}, I)$. Such problems of parameter estimation were the subject of Chapters 2 and 3; with the usual approximations for the prior and likelihood, they frequently reduce to a least-squares analysis.

The next situation arises when we don't know enough about $f(x)$ to be able to specify a particular functional model unambiguously, but can restrict ourselves to a small set of possibilities. A chemist concerned with studying the rotational excitations of a molecular compound, for example, could use quantum mechanical arguments to decide that the spectrum of interest should consist of a few sharp peaks; nevertheless, there may be some uncertainty as to their expected number. For illustrative purposes, eqn (6.1) could then be generalized so that $f(x)$ is defined by a sum like

$$f(x) = \sum_{j=1}^{M} a_j \, \delta \left(x - x_j \right), \tag{6.2}$$

where it is only known that M is a small integer ($\leqslant 10$, say). The problem of estimating the optimal value of M requires the evaluation of the posterior pdf $\mathrm{prob}(M \,|\, \{D_k\}, I)$; this task, and model selection in general, was the subject of Chapter 4.

The most difficult case occurs when we know so little about the object of interest that we are unable to characterize it by a functional model. This is a rather awkward situation because we cannot communicate quantitatively about $f(x)$ without some form of parametric description. Given this limitation, we must try to choose a formulation that ensures a lot of flexibility in the allowed solutions. This usually means that $f(x)$ has to be defined by a very large number of variables, which leads to problems of its own. In particular, a naïve uniform prior pdf is no longer adequate when there are more parameters to be inferred than data. To take a trivial example, there is no unique best estimate for X and Y when we are only told that $X + 2Y = 3$; the ambiguity would be reduced to a finite interval if we knew that both X and Y were positive, and could be eliminated completely if we had reason to believe that they tended to be of the same magnitude. Thus, in order to obtain a satisfactory free-form solution, we are forced to think about any weak constraints that might be appropriate for $f(x)$. These can either be encoded through the use of a suitable non-uniform prior or implemented by a judicious choice of the description of $f(x)$, or a combination of the two; the bulk of this chapter is devoted to illustrating how this can be done. Before that, however, let us examine a simple 'non-parametric' set-up more closely to get a better appreciation of the difficulties associated with having too few data.

6.1.1   Singular value decomposition

Perhaps the most straightforward way of obtaining a free-form description of a one-dimensional spectrum $f(x)$ is to divide the x-axis into tiny pieces and consider the (average) value of f in each of these regions. This procedure is shown schematically in Fig. 6.1, and is equivalent to expressing $f(x)$ as a sum of many, slightly displaced, narrow rectangles. Ignoring the finite width of these columns, $f(x)$ is defined mathematically as in eqn (6.2), except that M is now large and the locations of the δ-functions $\{x_j\}$ are no longer unknown variables; they are fixed by a relationship such as

$$x_j = x_{\min} + \left[\frac{x_{\max} - x_{\min}}{M} \right] \left(j - \tfrac{1}{2} \right), \tag{6.3}$$

for $j = 1, 2, \ldots, M$. Our inference about $f(x)$ is, therefore, encapsulated in the posterior pdf for the amplitude parameters: $\mathrm{prob}\left(\{a_j\} \,|\, \{D_k\}, I \right)$. This division of a continuous function into a set of discrete *pixels* can be generalized to two, three and higher

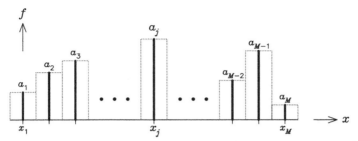

Fig. 6.1 A schematic illustration of the way a function $f(x)$ can be described by M amplitudes $\{a_j\}$ of a set of discrete pixels centred at $\{x_j\}$.

dimensions: we just have (rectangular) elements of area and volume, instead of length. In fact, this discretization is nothing more than the standard brute force and ignorance way of doing calculations numerically on a computer.

Having set up a free-form description in Fig. 6.1, let's turn our attention to the characteristics of the likelihood function $\text{prob}(\{D_k\}|\{a_j\}, I)$. If this has a well-defined maximum, when viewed in the M-dimensional space of the $\{a_j\}$, then a naïve uniform prior will be sufficient to yield a unique best estimate for $f(x)$; if it does not, then we will have to think harder about a suitable assignment for $\text{prob}(\{a_j\}|I)$. To keep the analysis simple, let us work with the $\exp(-\chi^2/2)$ likelihood function of Section 3.5:

$$\log_e\left[\text{prob}(\{D_k\}|\{a_j\}, I)\right] = \text{constant} - \frac{1}{2}\sum_{k=1}^{N}\left(\frac{F_k - D_k}{\sigma_k}\right)^2, \qquad (6.4)$$

where σ_k is the (known) error-bar for the kth datum D_k, and F_k is the corresponding prediction based on the given amplitude parameters $\{a_j\}$. We will also take it that the data are related linearly to $f(x)$ so that F_k can be written as

$$F_k = \sum_{j=1}^{M} T_{kj}\, a_j + C_k\,, \qquad (6.5)$$

where the elements of the matrix $\mathbf{T}$ and vector C are determined by the nature of the experiment but are independent of the values of the $\{a_j\}$. For example, T_{kj} might be equal to the sine, or cosine, of $2\pi jk/M$ for interferometric measurements and C could represent a slowly varying background signal.

Within the congenial conditions provided by eqns (6.4) and (6.5), the shape of the likelihood function is elliptical in nature. As such, the illustration of Fig. 3.6 represents the special case of $M = 2$ (with $X = a_1$ and $Y = a_2$). Although ellipsoids of higher dimensions are difficult to draw, they too are characterized completely by the directions and widths of their principal axes. These are formally given by the eigenvectors $\{e_l\}$ and eigenvalues $\{\lambda_l\}$, respectively, of the $M \times M$ matrix of second-derivatives of $\log_e\left[\text{prob}(\{D_k\}|\{a_j\}, I)\right]$; ignoring a factor of $-1/2$, the latter is just $\nabla\nabla\chi^2$. Hence, the orientation and spread of the likelihood function can be ascertained by (numerically) solving the simultaneous equations

$$\nabla\nabla\chi^2 \, e_l \; = \; \lambda_l \, e_l \, , \qquad\qquad (6.6)$$

for $l = 1, 2, \ldots, M$, where the ijth element of the Hessian matrix is given by

$$\left[\nabla\nabla\chi^2\right]_{ij} \; = \; \frac{\partial^2\chi^2}{\partial a_i\,\partial a_j} \; = \; 2\sum_{k=1}^{N}\frac{T_{ki}\,T_{kj}}{\sigma_k^2}\,. \qquad\qquad (6.7)$$

Since the width of the probability ellipsoid along its principal axes is inversely proportional to the (square-root of the) corresponding eigenvalues, we would ideally like all the λ's to be large. By the same token, the presence of λ's with small values warns us that the maximum of the likelihood function is very shallow. Indeed, it is not defined (uniquely) if any of the eigenvalues is equal to zero, because there are then some directions in which $\mathrm{prob}\big(\{D_k\}\,|\,\{a_j\}, I\big)$ is absolutely flat. As it can be shown that this situation necessarily arises when the number of parameters to be inferred is greater that the number of data ($M > N$), it follows that a uniform prior pdf is not generally adequate for obtaining an unambiguous best solution for $f(x)$ in that case.

The occurrence of extremely small eigenvalues ($\lambda \sim 0$) is really the hallmark of the free-form solution problem, and can arise even when the number of data is large enough that $M \leqslant N$. The easiest way to check for them is to test whether the determinant of the $\nabla\nabla\chi^2$ matrix is (almost) equal to zero, because the latter is given by the product of the $\{\lambda_l\}$. Since a matrix whose determinant is zero is said to be *singular*, an analysis of its eigenproperties is called *singular value decomposition*. This can be a useful exercise as it enables us to ascertain which aspects of $f(x)$ can, or cannot, be inferred reliably from the data; let us elaborate on this further.

Consider the two-dimensional example of Fig. 3.6, with parameters X and Y. If the principal directions of the ellipse lay along the coordinate axes, then the related eigenvectors would be: $e_1 = (1, 0)$ and $e_2 = (0 , 1)$. The corresponding widths of the pdf, which are proportional to the reciprocal of $\sqrt{\lambda_1}$ and $\sqrt{\lambda_2}$, would clearly indicate the uncertainties associated with our best estimates of X and Y respectively. Now suppose that the ellipse is skew, as drawn in Fig. 3.6. The description of the reliability of the optimal inference is then not quite so straightforward, because the estimates of X and Y are correlated. For example, the data may tell us a lot about the sum $X + Y$ but almost nothing about the difference $X - Y$; in that case, the marginal error-bars σ_X and σ_Y will both be huge but the sum-constraint means that typical estimates X and Y will not be independent of each other. While this behaviour can be conveyed through a covariance matrix, as noted in Section 3.2, the same information is also contained in the eigenvectors and eigenvalues. This is because, for such a negative correlation, $e_1 = (1, -1)$ and $\lambda_1 \sim 0$, indicating that the difference $X - Y$ is poorly determined, and $e_2 = (1, 1)$ and λ_2 is large, showing that the sum $X + Y$ can be estimated with a high degree of confidence. More generally, an eigenvector with components (α, β) refers to the quantity $\alpha X + \beta Y$; the corresponding eigenvalue tells us how reliably it can be inferred. For our multivariate case, the M elements of the various e_l's pertain to different linear combinations of the $\{a_j\}$; since the principal axes of an ellipse are mutually perpendicular, these are the quantities which are uncorrelated and can be estimated independently of each other.

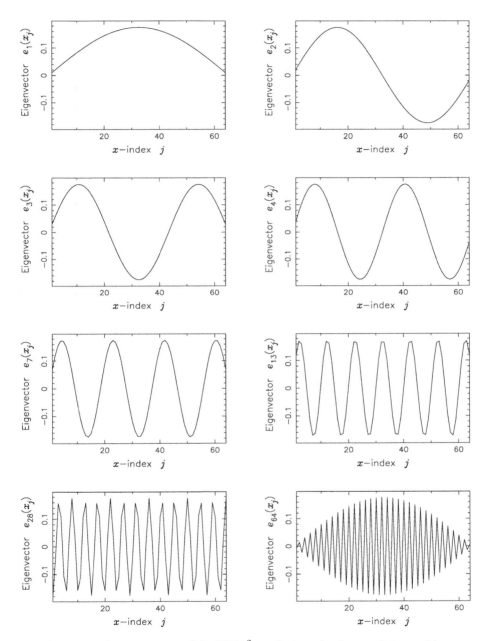

Fig. 6.2 Some of the eigenvectors of the $\nabla\nabla\chi^2$ matrix for a simple convolution problem, presented in order of decreasing eigenvalues; the calculation was done on a grid of 64 points. Rather than using the discrete 'stick' representation of Fig. 6.1, the vertices of adjacent amplitudes have been joined together to give a continuum impression.

Hence, the eigenvectors with large λ's represent aspects of $f(x)$ which can be inferred reliably from the data; those with small λ's indicate areas of great uncertainty.

As a concrete example of a singular value analysis, let's look at a convolution problem like the one illustrated in Fig. 4.2; in such a case, the experimental data are a blurred and noisy version of the object of interest. If the shape of the resolution (or blurring) function was Gaussian, with width w, then the elements of the **T** matrix in eqn (6.5) would take the form

$$ T_{kj} \;=\; \frac{1}{w\sqrt{2\pi}}\; \exp\left[-\,\frac{\left(x_k - x_j\right)^2}{2\,w^2} \right], $$

where x_k and x_j are the x-coordinates of the kth datum and jth pixel in the free-form representation of $f(x)$, respectively. Taking both the $\{a_j\}$ and the $\{D_k\}$ to be defined on the same (uniform) grid of 64 points, so that $M = N = 64$, and all the error-bars in eqn (6.7) to be unity ($\sigma_k = 1$), some of the eigenvectors of the corresponding $\nabla\nabla\chi^2$ matrix are shown in Fig. 6.2; they have been ordered by the magnitude of their eigenvalues, with $\lambda_1 > \lambda_2 > \lambda_3$ and so on. Each of the e_l's is, of course, a collection of 64 numbers: the components of a vector along the M parameter-axes, referring to a particular linear combination of the $\{a_j\}$. Rather than representing them by a set of discrete amplitudes for the various pixels, as in Fig. 6.1, adjacent vertices have been joined together to give a continuum impression in Fig. 6.2. Although the results plotted are for $w = 1$ pixel, the eigenvectors do not change very much with the blur width. In addition to their general oscillatory nature, the most important feature of the e_l's is that those associated with large eigenvalues vary slowly with x while the ones having small λ's are highly structured. This indicates that it is the gross, or low-frequency, features of $f(x)$ which can be inferred most reliably and emphasizes that information about the fine detail is lost in a convolution process. The spectra of eigenvalues for $w = 1$ and $w = 5$ pixels are shown in Fig. 6.3; since the latter has many more small λ's, it confirms our expectation that the problem becomes more acute as the blur width increases.

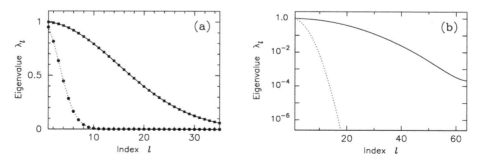

Fig. 6.3 The eigenvalue spectrum $\{\lambda_l\}$, for the eigenvectors in Fig. 6.2, plotted on: (a) a linear scale; (b) a logarithmic scale. The squares and solid line are for a blur-width of $w = 1$ pixel; the circles and dotted line are for $w = 5$ pixels.

6.1.2 A parametric free-form solution?

From our discussion so far, we have seen that a non-parametric solution for $f(x)$ actually entails the inference of a very large number of parameters. Although this does not pose any problem of principle, it does mean that we have to think much harder about a suitable non-uniform prior in order to obtain a satisfactory solution. An easy way of avoiding this inconvenience is to use a free-form description which uses fewer variables. We have already met one method of doing this in the guise of the Taylor series expansion, whereby a potentially complicated function is approximated (locally) by a low-order polynomial:

$$f(x) = c_1 + c_2(x-x_0) + c_3(x-x_0)^2 + c_4(x-x_0)^3 + \cdots . \qquad (6.8)$$

As long as the number of coefficients $\{c_l\}$ used is small enough, their estimation frequently reduces to a straightforward exercise in least-squares fitting; if necessary, we could carry out a model selection analysis to determine the optimal expansion order (as in Gull 1988). Even though a power series may not provide the best approximation for the spectrum of interest, eqn (6.8) illustrates how an arbitrary function can be expressed as a linear combination of M_b *basis functions* $\{\eta_l(x)\}$:

$$f(x) = \sum_{l=1}^{M_b} c_l \, \eta_l(x) . \qquad (6.9)$$

The Taylor series, therefore, represents the special case $\eta_l(x) = (x-x_0)^{l-1}$. Depending on the nature and the symmetry of the problem, *Chebyshev polynomials, spherical harmonics, sinusoids, wavelets* and many other functions have been found to be useful. The description of $f(x)$ in eqns (6.2) and (6.3) is also a special case of eqn (6.9), with $\eta_l(x) = \delta(x-x_l)$; the trick here, however, is to choose a set of basis functions carefully so that M_b can be made small. Although the aim is to reduce the analysis to simple parameter estimation, it can still be considered a free-from solution in the sense that the coefficients $\{c_l\}$ don't directly correspond to meaningful quantities in terms of a particular physical model.

In many ways, the eigenvectors of the $\nabla\nabla\chi^2$ matrix provide a natural choice for the expansion of eqn (6.9); in the continuum limit, they are called *eigenfunctions*: $\eta_l(x) = e_l(x)$. Those associated with large eigenvalues represent aspects of $f(x)$ which are determined well by the experimental measurements, while the ones with small λ's make almost no contribution towards the fitting of the data; this insensitivity of the latter means that they can be omitted from the summation. In terms of our discrete formulation of Fig. 6.1, the spectrum of interest then becomes

$$f(x_j) = a_j = \sum_{l=1}^{M_b} c_l \, e_l(x_j) , \qquad (6.10)$$

for $j = 1, 2, \ldots, M$, where $e_l(x_j)$ is the jth component of the lth eigenvector in eqn (6.6); the sum is only over those e_l's that have large λ_l's, so that typically $M_b \ll M$.

After some effort, it can be shown that the optimality condition $\nabla\chi^2 = 0$ and the covariance matrix $2\left(\nabla\nabla\chi^2\right)^{-1}$ lead to the result

$$c_l = \frac{2}{\lambda_l} \sum_{k=1}^{N} \frac{D_k}{\sigma_k^2} \sum_{j=1}^{M} T_{kj}\, e_l(x_j) \pm \sqrt{\frac{2}{\lambda_l}}. \qquad (6.11)$$

In the derivation of this formula for the coefficients $\{c_l\}$, it is useful to remember that the eigenvectors of a real symmetric matrix (like the Hessian $\nabla\nabla\chi^2$) are *orthogonal*; by convention, they are also normalized so that

$$\sum_{j=1}^{M} e_l(x_j)\, e_m(x_j) = \begin{cases} 1 & \text{if } l = m, \\ 0 & \text{otherwise}. \end{cases}$$

Although a singular value decomposition analysis allows us to circumvent the technical difficulties of handling very small eigenvalues, it does not really address the central issue of the infinite uncertainties inherent in seeking a free-form solution. This is because the addition of any multiple of an eigenvector with $\lambda = 0$ in eqn (6.10) changes the estimate of $f(x)$, but not the fit to the data $\mathrm{prob}[\{D_k\}|f(x), I]$. To break this ambiguity, and have a unique maximum for the posterior pdf $\mathrm{prob}[f(x)|\{D_k\}, I]$, we must ponder over whatever little information I contains about $f(x)$ and assign an appropriate non-uniform prior $\mathrm{prob}[f(x)|I]$.

6.2   MaxEnt: images, monkeys and a non-uniform prior

Suppose we knew that the object of interest constituted a positive and additive distribution, and almost nothing else. By this we mean that $f(x) \geqslant 0$ and its integral (or sum) over any region represents a physically meaningful and important quantity. This is so for the radio-flux distribution of an astronomical source, the electron density of a crystal, the intensity (but not amplitude) of incoherent light in an optical image and many other situations. What prior best encodes our limited state of knowledge?

Well, since the properties under discussion also hold for pdfs, the principle of maximum entropy encountered in Chapter 5 may be helpful to us here as well. In particular, an easy way of thinking about distributions that are positive and additive is to imagine that they are the result of a team of monkeys scattering balls of 'stuff' at random into a collection of boxes; if the latter are arranged in a two-dimensional grid, and the little blobs being thrown are luminescent, then we could generically call the end product an image. While no distribution of interest is actually generated in this manner, the mechanism is a good representation of our state of ignorance. As such, it seems reasonable that we should assign the highest (initial) probability to those $f(x)$'s which our hypothetical apes come up with most often, and the least to the ones that are rarest. Hence, following the discrete formulation in Fig. 6.1, eqn (5.28) suggests that a suitable prior pdf might be

$$\mathrm{prob}(\{a_j\}|\{m_j\}, \alpha, I) \propto \exp(\alpha S), \qquad (6.12)$$

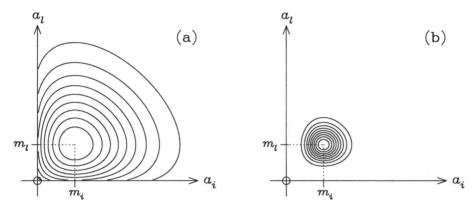

Fig. 6.4 The contours along which $\exp(\alpha S)$ is constant, as a function of the amplitude a_i and a_l in two pixels i and l; the Lebesgue measure is uniform, so that $m_i = m_l$. In (a) $\alpha = 1$, while in (b) $\alpha = 10$.

where α is a constant and S is the entropy of the $\{a_j\}$ relative to the Lebesgue measure $\{m_j\}$. According to Skilling (1989, 1991), the Shannon–Jaynes expression for S should be generalized to

$$S = \sum_{j=1}^{M} \left(a_j - m_j - a_j \log\left[\frac{a_j}{m_j}\right] \right), \tag{6.13}$$

if the distributions involved are not normalized; it reduces to the form in eqn (5.28) when $\sum a_j = \sum m_j = 1$, and to the even more basic $-\sum a_j \log(a_j)$ formula of eqn (5.26) if the measure is also taken as being uniform (so that $m_j = 1/M$). The pdf of eqns (6.12) and (6.13) is illustrated in Fig. 6.4. It has a maximum at $\{a_j\} = \{m_j\}$, and decreases rapidly to zero as any of the $\{a_j\}$ tries to become negative. The width of the pdf is controlled by the parameter α which, in terms of the analysis of Section 5.2, is equivalent to the total number of balls (or pennies etc.) thrown by the monkeys: the larger the value of α, the sharper the entropic prior.

Using the assignment in eqn (6.12), and the least-squares likelihood function of eqn (6.4), we obtain the posterior pdf:

$$\mathrm{prob}\big(\{a_j\}|\{D_k\},\{m_j\},\alpha,I\big) \propto \exp\left[\alpha S - \frac{\chi^2}{2}\right]. \tag{6.14}$$

This follows from Bayes' theorem, with α and $\{m_j\}$ given throughout, as the data depend only on the $\{a_j\}$: $\mathrm{prob}\big(\{D_k\}|\{a_j\},\{m_j\},\alpha,I\big) = \mathrm{prob}\big(\{D_k\}|\{a_j\},I\big)$. With the extraneous factors in the conditioning statement, this isn't quite the pdf we were after! Although Gull (1989a) tried to deal with α properly through marginalization, Sibisi (1996) and Skilling (1998) have concluded that the MaxEnt prior of eqn (6.12) has an inherent flaw: the results are dependent of the degree of pixelation (or choice of M). This shortcoming had been missed earlier due to a deceptive side-effect of the Gaussian approximation made in the calculation, and because the quantitative answers from the analysis (Skilling 1990) were generally sensible in practice (Sibisi 1990, Sivia and Webster 1998). Rather than abandoning the idea of MaxEnt image processing altogether, we can make use of its helpful properties by adopting the pragmatic, if qualitative, approach of *regularization* for this difficult problem; a probabilistic formulation is discussed in Chapter 10, where more advanced computational techniques are described.

6.2.1 Regularization

Since the 'most probable' estimate for the $\{a_j\}$ in eqn (6.14) is given by the maximum of $\alpha S - \chi^2/2$, this condition can be regarded as a constrained minimization of χ^2 where α is a Lagrange multiplier. From this viewpoint, S is seen as a regularization function which helps to stabilize the least-squares procedure for a free-form solution; many alternatives have been proposed for this purpose (Tikhonov and Arsenin 1977). The most popular choice is, perhaps, the quadratic penalty $\sum (a_j - m_j)^2$, which offers the advantages of analytical simplicity; others include nearest-neighbour smoothness criteria, such as $\sum (a_{j+1} - a_j)^2$, and the Burg (1967) entropy $\sum \log(a_j/m_j)$. Within a probabilistic framework, the various functions would be equivalent to different non-uniform priors with each reflecting whatever limited information there was available. Some of the earliest examples of image processing with MaxEnt regularization were in astronomy (Frieden 1972, Gull and Daniell 1978), and a classic case from a forensic study is reproduced in Fig. 6.5 (Gull and Skilling 1984).

The task of solving the large set of simultaneous equations

$$\frac{\partial}{\partial a_j} \left(\alpha S - \frac{\chi^2}{2} \right) = 0 \,, \qquad (6.15)$$

for $j = 1, 2, \ldots, M$, can be a demanding numerical exercise. This is particularly so if the number of variables involved is much greater than a couple of hundred, since the

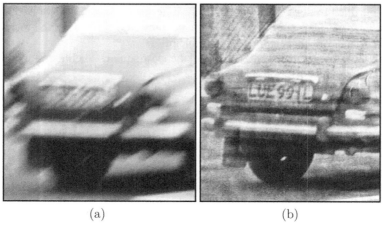

(a)   (b)

Fig. 6.5 (a) Data consisting of a photograph, where the picture of interest has been blurred by the movement of the camera. (b) The MaxEnt estimate of the underlying scene.

computation time for matrix manipulations tends to scale as the third power of the size of the problem ($\sim M^3$). Skilling and Bryan (1984) describe a good general-purpose algorithm for this optimization, with an elaborate iterative scheme at its core. The procedure begins at the global maximum of entropy, $\{a_j\} = \{m_j\}$, which is the estimate of $f(x)$ returned in the absence of any data; in this context, the Lebesgue measure is often called the *default model*. The local gradients of S and χ^2 are then consulted, to ascertain the best small change $\{\Delta a_j\}$ which will reduce the misfit with the data while keeping the entropy as large as possible. This is done repeatedly until the desired solution is obtained, $2\alpha \boldsymbol{\nabla} S = \boldsymbol{\nabla} \chi^2$, with $\chi^2 \approx N$ (the number of measurements) to ensure a respectable agreement with the data; it is illustrated schematically in Fig. 6.6.

There are two important conditions that must be satisfied if the algorithm outlined above is to converge satisfactorily. The first is that the likelihood function must not be multimodal. If it is, then the posterior pdf will inherit the unfortunate aspects of that topology; and all the grief it entails (c.f. Sections 3.2 and 3.4)! A notorious example of this situation concerns the case of *phaseless* Fourier data which bedevils crystallographers, amongst others. Luckily, the likelihood function for many problems is well-behaved, as it can be characterized by eqns (6.4) and (6.5). This is implicitly assumed in Fig. 6.6, because the contours of χ^2 have been drawn as being elliptical. Given the *convex* nature of both S and χ^2 under these circumstances, so that they can only touch each other at a single point, the MaxEnt solution is unique and 'easy' to find.

The second condition for satisfactory convergence stems from a more practical consideration. It concerns the maximum change in the image $\{\Delta a_j\}$ allowed between one iteration and the next. This is because the local gradients of S and χ^2, upon which the update is based, are a poor guide to the long-range behaviour of $\alpha S - \chi^2/2$. In order to impose a limit on the size of the permissible change, we must first decide on how the 'step-length' Δl is to be defined. Skilling and Bryan (1984) have found that a good

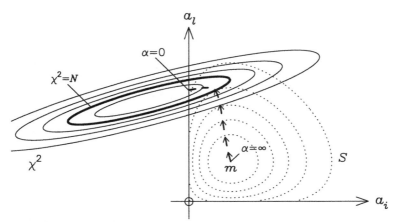

Fig. 6.6 A schematic illustration of a MaxEnt trajectory. We start at the default model m, and take small steps in such a way as to reduce the misfit with the data (χ^2) while keeping the entropy S as large as possible.

choice is

$$\Delta l^2 = \sum_{j=1}^{M} \frac{\Delta a_j^2}{a_j}, \tag{6.16}$$

where Δa_j is the small increment in the amplitude of the jth pixel. A little reflection soon reveals the benefits of this non-Euclidean choice, with the extra a_j in the denominator: a small change in a pixel with a low flux (e.g. $\Delta a_j = 0.1$ for $a_j = 1$) is comparable to a large deviation for one with a big amplitude (e.g. $\Delta a_j = 1$ for $a_j = 100$). This property is useful for avoiding the inadvertent development of negative amplitudes. Although higher powers of a_j in the denominator would also have this effect, they tend to be unnecessarily conservative and, hence, less efficient.

For some types of problems, such as those where the object of interest is related to the data through a *Laplace transform*, the general-purpose algorithm outlined above is not very efficient. Bryan (1990) has developed an alternative procedure specifically for these cases, where the likelihood function only has a handful of good eigenvalues.

6.3 Smoothness: fuzzy pixels and spatial correlations

Maximum entropy regularization is a good way of encoding our prior knowledge that $f(x)$ is positive quantity, whose integral represents the amount of 'important stuff' in any region. Often, however, we have reason to believe that the object of interest is also locally smooth. How can we build this into the analysis? This question may seem strange at first, since MaxEnt images are frequently described as being the smoothest solutions consistent with the data. While it is true that our hypothetical team of monkeys will generate featureless maps far more often than highly structured ones, this is not quite the same thing as local smoothness. In some ways, the concepts are incompatible: the entropic formula arises from saying that the next ball is equally likely to be dropped into any bin, irrespective of the number that are already there; smoothness implies that the structure is spatially correlated, so that a high flux in one pixel is accompanied by comparably large values in its neighbours. How can these conflicting requirements be accommodated?

A simple resolution presents itself if we return to Fig. 6.1. There we said that an arbitrary function $f(x)$ could be expressed as a sum of many, slightly-displaced, δ-functions (or extremely narrow rectangles); in terms of eqn (6.9), this corresponds to using a set of basis functions $\eta_j(x) = \delta(x - x_j)$. If we want to describe a smooth function, in a free-form manner, all we need to do is use a set of basis functions which have some spatial extent. Using the familiar Gaussian form of width w, for example, eqn (6.2) would then become

$$f(x) = \sum_{j=1}^{M} a_j \exp\left[-\frac{(x - x_j)^2}{2 w^2} \right]. \tag{6.17}$$

This situation is illustrated in Fig. 6.7. For any given length-scale w, the problem reduces to the one of estimating the positive amplitudes $\{a_j\}$ that was met in the previous section. Thus 'fuzzy pixels' (or fuzzy balls) provide an easy extension to the monkey model which allows us to encode preference for local smoothness.

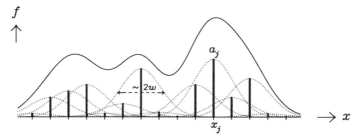

Fig. 6.7 A schematic illustration of the way a smooth function $f(x)$ can be represented as the sum of basis functions with spatial extent w.

There are two obvious questions for the above formulation: how do we choose w, and do we have to use this specific basis function setup? Within the regularization framework, the pragmatic advice is to let w be as large as possible while still satisfying the constraints of the data. This is a qualitative way of implementing Ockham's razor: if w is too small, unnecessary structure will become admissible; if w is too big, $\chi^2 \approx N$ will not be achievable. The answer to the second question is no: the use of Gaussian basis functions, of fixed width, is just a simple choice that is often found to be adequate; if it isn't, we'd have to try something a bit more elaborate.

Although the fuzzy pixel formulation and more traditional gradient constraints, where $\sum (a_{j+1} - a_j)^2$ might be used as a regularizer, are essentially alternative ways of thinking about local smoothness, the former does offer some advantages. First, it makes it easy to combine smoothness with positivity: as long as both the basis functions and their amplitudes are positive, this will also be true of $f(x)$. Second, since the existence of a derivative implies continuity, the use of gradient constraints is technically restricted to such functions. By contrast, the basis functions of Fig. 6.7 can be either continuous or discrete; as such, this idea of smoothness as a correlation between spatially-related pixels is far more general. Indeed, this brings us to the last point: unlike derivatives, the concept of fuzzy pixels can easily be extended to encode prior knowledge about the non-local behaviour of $f(x)$. For example, in a stellar spectrum, we might expect that an emission line at frequency ν_1 would be accompanied by a similar companion at $2\nu_1$; in that situation, we could use basis functions which had two peaks separated by a suitably large distance.

6.3.1 Interpolation

The importance of using prior knowledge about local smoothness is, perhaps, most vividly demonstrated by the *interpolation* problem. Here the data consist of the actual values of $f(x)$ given at coarse intervals; we are then required to make inferences about this function on a much finer scale. An example of such data, with imperceptibly small (but finite) error-bars, is shown in Fig. 6.8(a). If we only know that $f(x)$ is a positive distribution, and maximize the entropy as in Section 6.2, then we will obtain the solution of Fig. 6.8(b). This answer looks extremely weird, but is easily explained. The data tell us a lot about $f(x)$ at the measured points, but they contain absolutely no information

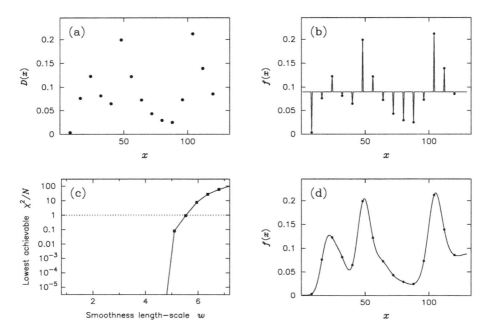

Fig. 6.8 (a) Interpolation data. (b) The MaxEnt solution of Section 6.2, equivalent to $w=0$. (c) A plot of the lowest achievable value of the normalized χ^2 versus the spatial correlation length-scale w. (d) The MaxEnt solution with $w=4$.

about its behaviour in the intervening space. Consequently, while the solution is highly constrained at the data-points, the monkeys are free to even-out everything in between. Hence, the very spiky appearance of Fig. 6.8(b).

Although understandable, the result in Fig. 6.8(b) is still disconcerting. The cause of this unease is, of course, the lack of local smoothness. If we encode our prior expectation of spatial correlations through the fuzzy pixel formulation of eqn (6.17), then we obtain the graph of the lowest achievable χ^2/N versus w shown in Fig. 6.8(c); values of w less than about five are capable of giving adequate agreement with the data ($\chi^2 \approx N$). The shape of the curve in Fig. 6.8(c) is highly atypical, being peculiar to the interpolation problem; usually, it's not easy to achieve $\chi^2 \ll N$ even as $w \rightarrow 0$. The MaxEnt solution with $w=4$ is given in Fig. 6.8(d).

6.4   Generalizations: some extensions and comments

In Section 6.2, we said that the entropic regularizer of eqn (6.13) was appropriate for the case where it was only known that $f(x)$ constituted a positive and additive distribution. Although this is quite common, what can be done if (even) these conditions are not valid? Well, we obviously need to use a regularization function for the $\{a_j\}$ that does not exclude negative values; to overcome the non-uniqueness difficulties of free-form solutions, it's still helpful if it has a well-defined maximum. A quadratic penalty factor is a suitable candidate, and can be thought of as a multivariate Gaussian prior:

$$\text{prob}(\{a_j\}|\{m_j\},\{\gamma_j\},\alpha,I) \propto \exp\left[-\alpha \sum_{j=1}^{M} \frac{(a_j-m_j)^2}{\gamma_j^2}\right]. \qquad (6.18)$$

The logarithm of this expression simplifies to $-\alpha \sum (a_j - m_j)^2$ if all the γ_j are equal, and to just $-\alpha \sum a_j^2$ if all the m_j are also zero. Not only is eqn (6.18) less constraining than its entropic counterpart, it is much easier to handle analytically. When α is large and all the m_j are positive, the Gaussian and entropic priors are very similar; this can be verified visually from Fig. 6.4(b), or derived formally by expanding the entropy of eqn (6.13) as a (multivariate) quadratic Taylor series about its maximum at $\{a_j\}=\{m_j\}$:

$$S \approx -\sum_{j=1}^{M} \frac{(a_j-m_j)^2}{2m_j},$$

so that $\exp(\alpha S)$ is just like eqn (6.18), with $\gamma_j = \sqrt{2m_j}$. Thus, the simpler regularization function $\sum (a_j - m)^2$ can often lead to results which closely resemble the entropic ones. In accordance with its implied extra prior knowledge, the latter will be superior if positivity is an important (rather than merely valid) constraint; this will be particularly so if the object of interest has a large *dynamic range* (i.e. it has some very bright areas on generally dim background).

Even for the case when $f(x)$ is not necessarily positive, we might have cogent prior information to assign a more elaborate pdf than eqn (6.18). One such situation arises when the object of interest is known to be the difference of two distributions which are each inherently positive and additive. For example, the nuclear-spin density in an NMR experiment results from the combined affect of atoms with up and down spin; denoting their distributions by $f_u(x)$ and $f_d(x)$, respectively, $f(x) = f_u(x) - f_d(x)$. Our inference about $f(x)$ is, therefore, governed by what we can say about $f_u(x)$ and $f_d(x)$. Given only the most basic characteristics of the up and down distributions, it is appropriate to use a regularizer constructed from the sum of their individual entropies: $S = S_u + S_d$, where S_u and S_d are given by expressions similar to eqn (6.13). Since our primary interest tends to lie in $f(x)$, rather than $f_u(x)$ and $f_d(x)$, it is more convenient to work with it directly. This can be done, following some mathematical inspiration, if there is a common default model. It relies on the fact that the maximum of the total entropy S_u+S_d, subject to a constraint on any function of $f(x)$, or the data, is equivalent to the optimization of S_{ud}:

$$S_{ud} = \sum_{j=1}^{M} \left(\psi_j - 2m_j - a_j \log\left[\frac{\psi_j + a_j}{2m_j}\right]\right),$$

where $\psi_j^2 = a_j^2 + 4m_j^2$, and a_j is not necessarily positive. The use of this regularizer, and that of eqn (6.18), is illustrated in Fig. 6.9; the data consist of the 18 lowest-order Fourier coefficients of a 'spin-density' function defined on a grid of 128 points. We see that, by comparison with the simpler Gaussian prior, the best estimate of $f(x)$ from the

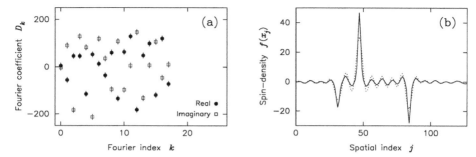

Fig. 6.9 (a) Simulated data, resulting from the Fourier transform of a 'spin-density' function. (b) The corresponding best estimates of $f(x)$ obtained by using a hybrid positive–negative entropy formulation (solid line) and a Gaussian prior (dotted line).

hybrid positive–negative entropy formulation has somewhat reduced rippling features and sharper 'structural' peaks. Two examples of its practical use can be found in Laue *et al.* (1985) and David (1990).

Returning to the case where the object of interest is positive, suppose we also knew it had an upper bound f_{max}. How can we take this additional prior information into account? Well, we obviously need a regularizer which is zero unless $0 \leqslant f(x) \leqslant f_{max}$; for a unique (best) free-form solution, it's also helpful if it has a well-defined maximum. One such possibility is

$$ S = - \sum_{j=1}^{M} \left(a_j \log\left[\frac{a_j}{m_j}\right] + (1-a_j) \log\left[\frac{1-a_j}{1-m_j}\right] \right), \tag{6.19} $$

where we have set $a_{max} = 1$; it has a maximum at $\{a_j\} = \{m_j\}$, and decays rapidly to zero as any of the amplitudes a_j approach 0 or 1. In fact, eqn (6.19) is also the sum of the entropies of two distributions: $\{a_j\}$ relative to the default model $\{m_j\}$, and $\{1-a_j\}$ relative to $\{1-m_j\}$.

6.4.1 Summary of the basic strategy

We've now seen several examples of how to deal with the free-form estimation of $f(x)$. In keeping with the spirit of flexibility, the procedures were designed to encode different types of weak assumptions about the object of interest: positivity, local smoothness, bounds and so on. Every formulation corresponds to a particular set of conditioning information and, in essence, represents a variation on the exact question that is being asked of the data; as such, data analysis is nothing more than a dialogue with the experimental measurements. If we don't 'like' the results of an analysis, then the constructive way forward is to reflect on the reasons for the dissatisfaction. This pondering usually reveals that we've either made an assumption which isn't valid or, as is more often the case, we haven't encoded a property of $f(x)$ we believe to be true. With this additional insight, the problem can then be recast so that we better answer the question which we had really intended to ask. The great advantage of this adaptive approach is that we don't have to fear 'failure': as long as we are guided to a more satisfactory analysis, we will have learnt something useful from the exercise. Indeed, it can be said that it's precisely when our initial expectations are not met, and we are forced to think more deeply about a problem, that real (scientific) progress is made.

Pursuing the theme a little further, we should emphasize that there is nothing to prevent us from carrying out a parametric analysis (as in Chapters 2 and 3) after a free-form estimation. As we have said before, the latter is used in situations where we know so little about the object of interest that we don't have the confidence to describe $f(x)$ by a specific functional model; its results, qualitative as they might be, can often suggest one! For example, Fig. 6.9(b) seems to indicate that the 'spin-density' function consists of just three sharp peaks, which could each be parameterized by an amplitude and a position. Not only will these stronger prior assumptions allow us to estimate $f(x)$ in a more succinct and precise manner, it will also automatically remove the unsightly rippling structure. We don't have to go all the way to committing ourselves to exactly three peaks, of course, but could simply say there were only a few; the model selection analysis of Chapter 4 could then be used to tell us how many there was most evidence for in the data. Although this transition from a free-from analysis to a model-based one is quite common, there is no reason why it shouldn't go the other way. This is particularly so if we have serious doubts about the validity of the (quite) strong prior knowledge which implicitly underlies any parametric estimation. In this case, we can carry out a free-form analysis to see whether our conclusions change very much as the initial constraints are relaxed; if they do, then we must lay more stress on the nature and importance of the related assumptions.

6.4.2   Inference or inversion?

In our discussion about free-form solutions at the beginning of this chapter, we largely restricted ourselves to the case where the data are related linearly to $f(x)$. Such problems are the ones which are most generally tractable, and can be written in the form of eqn (6.5); using matrix and vector notation, this becomes

$$F = \mathbf{T}\,a + C, \tag{6.20}$$

where a is a digitized version of $f(x)$, which gives rise to the ideal data F through an experimental set-up described by $\mathbf{T}$ and C. Given eqn (6.20), it is tempting to think of free-form analysis as an exercise in applying an *inverse operator* to the data:

$$a = \mathbf{T}^{-1}(F - C). \tag{6.21}$$

A little thought, however, soon reveals that this procedure is likely to be of limited use. First of all, the inverse only exists if $\mathbf{T}$ is a square matrix; since the number of pixels M in a is usually much larger than the number of data N, we are immediately in trouble. Secondly, eqn (6.21) is inadequate because it makes no mention of the noise σ which accompanies the experimental measurements. Finally, eqn (6.21) cannot be generalized easily for the case of non-linear problems.

Despite the drawbacks, the thought of being able to 'invert' the data directly can still hold considerable appeal; so, let's examine this idea a bit further with the aid of a simple convolution example. Suppose that the data D are related to $f(x)$ through a blurring with an invariant point-spread, or resolution, function $R(x)$:

$$D(x_k) = \int f(x)\, R(x_k - x)\, \mathrm{d}x \; + \; B(x_k) \; \pm \; \sigma(x_k)\,, \qquad (6.22)$$

for $k = 1, 2, \ldots, N$, where B is a slowly-varying background signal and σ is the noise in the measurements; in fact, we met this relationship in eqn (4.13) and Fig. 4.2. This equation cannot be inverted as it stands, but does become more amenable to analytical manipulation if we ignore σ and treat B and D like continuous functions. Then, according to the convolution theorem, the Fourier transform of eqn (6.22) yields

$$\hat{D}(\omega) \approx \hat{f}(\omega)\, \hat{R}(\omega) + \hat{B}(\omega)\,, \qquad (6.23)$$

where the (complex) hat-ω functions are related to their (real) x counterparts by

$$\hat{f}(\omega) = \int f(x)\, \exp(\mathrm{i}2\pi\omega x)\, \mathrm{d}x\,, \qquad (6.24)$$

and so on. By rearranging eqn (6.23), and then taking an inverse Fourier transform, the object of interest can be expressed as

$$f(x) \approx \int \left[\frac{\hat{D}(\omega) - \hat{B}(\omega)}{\hat{R}(\omega)} \right] \exp(-\mathrm{i}2\pi\omega x)\, \mathrm{d}\omega\,. \qquad (6.25)$$

The use of this inversion formula is illustrated in Fig. 6.10, where it is assumed that both B and R are known (exactly).

The data in Fig. 6.10(c) were simulated by convolving the test object of Fig. 6.10(a) with the resolution function shown in Fig. 6.10(b); a flat background signal and noise were also added. Since both $f(x)$ and $\{D_k\}$ are defined on the same grid of 128 points in this example, there is no difficulty with the existence of the equivalent inverse operator $\mathbf{T}^{-1}$. The result obtained by using eqn (6.25) is shown in Fig. 6.10(d); it provides an atrocious estimate of the test object of Fig. 6.10(a). Indeed, the blurred version of $f(x)$ given by the data is far more informative than the deconvolution! The cause of this disappointing performance is, of course, the failure of eqn (6.25) to take into account the presence of the noise in the measurements. Retaining the σ-term in eqn (6.22), a more accurate version of eqn (6.25) would be

$$\int \left[\frac{\hat{D}(\omega) - \hat{B}(\omega)}{\hat{R}(\omega)} \right] \exp(-\mathrm{i}2\pi\omega x)\, \mathrm{d}\omega \; = \; f(x) \pm \epsilon(x)\,, \qquad (6.26)$$

where the discrepancy ϵ is given by

$$\epsilon(x) \approx \int \left[\frac{\hat{\sigma}(\omega)}{\hat{R}(\omega)} \right] \exp(-\mathrm{i}2\pi\omega x)\, \mathrm{d}\omega\,. \qquad (6.27)$$

Hence, the error in the estimate of $f(x)$ is dominated by the small Fourier components of the resolution function; since the magnitude of $\hat{R}(\omega)$ tends to zero as ω becomes

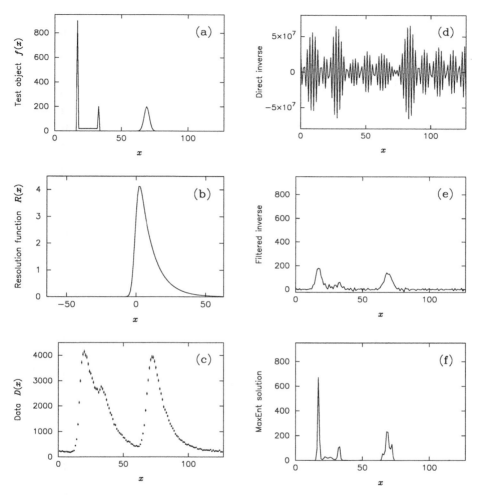

Fig. 6.10 The data shown in (c) were generated by convolving the test object of (a) with the resolution, or point-spread, function of (b); the simulation was carried out on a grid of 128 points, and a flat background signal and noise were also added. (d) is the result of using the 'inversion' formula of eqn (6.25), and (e) is the corresponding estimate of $f(x)$ given by a simple Gaussian filter. (f) is the MaxEnt solution.

large, most of the artefacts in Fig. 6.10(d) are high-frequency ripples. This suggests that the result of the inverse transform can be improved upon by smoothing it out,

$$f(x) \approx \int \left[\frac{\hat{D}(\omega) - \hat{B}(\omega)}{\hat{R}(\omega)} \right] \hat{\Omega}(\omega) \exp(-\mathrm{i}2\pi\omega x)\, \mathrm{d}\omega , \qquad (6.28)$$

where $\hat{\Omega}(\omega)$ is mostly near 1 but tends to 0 as least as fast as $\hat{R}(\omega)$ when $\hat{R}(\omega)$ is small, and the process is known as *filtering* or *windowing*. Carrying out such a blurring process, we obtain Fig. 6.10(e). This is clearly much better than Fig. 6.10(d): the long tails of the resolution function seen in Fig. 6.10(c) have been removed, but the 'true' features are no longer obscured by large-magnitude rippling artefacts.

The example of Fig. 6.10 illustrates the point that the inverse transform, even when it does exist, is usually not very helpful; additional procedures, such as filtering, are needed to make it usable. The crux of the problem is, of course, that data analysis is not a question of mathematical inversion: it's a matter of inference. The analysis that is required is the one that will carefully weigh-up the evidence of both the current data and our prior knowledge. The benefits of such an approach can be seen in Fig. 6.10(f), which shows the MaxEnt solution for $f(x)$. At least in this case, prior knowledge about positivity is a very powerful piece of information; consequently, it leads to a superior result.

Although data analysis, particularly free-form estimation, is often referred to as an 'inverse problem', we should not take this too literally. The task is really one of inferential calculus or plausible reasoning; while this is like the reverse of deductive logic, it's not the same as applying $\mathbf{T}^{-1}$ to the data. Having said that, a pseudo-inverse can be a practical tool for a qualitative analysis: it's certainly more computationally efficient than a constrained optimization, for example, and may well be adequate for the job at hand. Indeed, filtering can even be regarded as a kind of singular value decomposition technique because the effect of damping the artefact-prone high-ω components in eqn (6.28) is very similar to the omission of the small-λ eigenvectors in an expansion like eqn (6.10). As also noted in earlier chapters, therefore, many seemingly *ad hoc* (but popular) analysis procedures can be justified by reformulating them in terms of probability theory. This forces us to think more carefully about the assumptions which implicitly underlie them and, thereby, opens up the possibility for further improvement.

6.4.3   Advanced examples

The cases where there is uncertainty about the background signal or resolution function are more difficult, but progress can sometimes be made by adopting a multiple channel approach; we met two-channel entropy when discussing non-positive and bounded $f(x)$ a little earlier. A separation between background and signal can be attempted, for example, if a clear distinction can be made between broad and sharp structure (Yethiraj *et al.* 1991). Although an impressive illustration of 'blind deconvolution' (where $R(x)$ is unknown) can be found in Newton (1985), this problem does not generally have a satisfactory solution. Newton (1985) also has a somewhat exotic demonstration where a variety of (unknown) moving objects are automatically sorted into velocity-groups based on just two consecutive frames of a movie. A more practical application of multi-channel estimation, concerning the multi-scale analysis of data from the infra-red satellite IRAS, is reported in Bontekoe (1993).

An approach that has recently gained popularity is that of *independent component analysis*, or ICA, which has been used to tackle the 'blind source separation' problem. While details can be found in Roberts and Stephens (2001), for example, we recommend Knuth (1999, 2005) for a throughly Bayesian perspective on the topic.

7 Experimental design

We have now seen many examples of how probability theory can be used to make quantitative inferences based on all the (new) empirical information, and (past) prior knowledge, that is available. While this is the central task of data analysis, we have so far assumed that there is little choice in the actual conduct of the experiments. Let us turn our attention to this question, and consider how issues such as optimal instrumental design, and procedure, can be addressed.

7.1 Introduction: general issues

In Chapter 2, we began by considering the analysis of a simple coin-flipping experiment: Given that N tosses yielded R heads, what is the best estimate of the bias-weighting H and how confident are we in this prediction? Although the subsequent examples were of increasing technical complexity, they did not differ in their basic logical character; the problems were all to do with making quantitative inferences based on the relevant data, and prior knowledge, that were available. An important complementary issue, which we have not yet discussed, concerns the good design of experiments. This can include anything from an assessment of the number of measurements needed to achieve a desired level of confidence, to questions of the best way to build apparatus or collect data. Let us now see how probability theory can be used to guide us in such matters.

The simplest case is again provided by the coin-flipping example of Section 2.1: there is only one quantity to be inferred (the bias-weighting) and only one way of changing the experiment (the number of tosses); what's more, there are no nuisance parameters. The sole design question open is then: how many flips N will be required to estimate H to a given degree of confidence? Well, the reliability of the inference is described by the width of the posterior pdf $\mathrm{prob}(H\,|\,N, R, I)$; as we saw in Section 2.2, this can be approximated by the error-bar σ of eqn (2.20):

$$\sigma = \sqrt{\frac{H_\mathrm{O}(1-H_\mathrm{O})}{N}}\,, \tag{7.1}$$

where, according to eqn (2.19), $H_\mathrm{O} = R/N$. Hence, the confidence interval (a few times σ) depends on both the number of tosses and the fraction of heads obtained. While a few flips may be enough to convince us that a coin is heavily-weighted, a much larger number will be needed for us to be sure that it is reasonably fair. Thus we see that, even in this elementary case, the answer to our question is not so straightforward: it depends on the actual bias of the coin!

Although it's difficult to give a definitive answer for the number of flips needed for σ to be less than some small value σ_c, useful information can be still gleaned from eqn (7.1). For example, it tells us that the error-bar will be largest, for a given N, if $H_O = 1/2$. A plausible estimate will be provided, therefore, if the calculation is done on this basis: $N \geqslant 0.25/\sigma_c^2$. In fact, this estimate will not be very pessimistic as long as the coin is not heavily-biased (since we need $N \geqslant 0.16/\sigma_c^2$ if $H_O = 0.2$ or 0.8). Perhaps the most important feature apparent from eqn (7.1) is that the size of the error-bar is inversely proportional to $\sqrt{N}$; this can be verified pictorially from Fig. 2.1, and relies on the fact that the fraction of heads obtained does not vary a lot after a modest number of tosses. This means that if we evaluate σ on the basis of 50 flips (say), then we can calculate how many more will be needed to achieve the desired level of confidence; a four-fold increase will halve the error-bar, and so on.

Having illustrated the nature of experimental design questions with a very elementary example, let's consider the problem from a more general viewpoint. We are aiming to make the best possible inference about some object, or issue, of interest, based on all the information that is available. As we have seen in the preceding chapters, this usually entails the evaluation of the posterior pdf for a set of appropriate parameters: $\mathrm{prob}(\{X_j\}|\{D_k\}, I)$. According to Bayes' theorem, this can be expressed as a product of the likelihood and prior:

$$\mathrm{prob}(\{X_j\}|\{D_k\}, I) \propto \mathrm{prob}(\{D_k\}|\{X_j\}, I) \times \mathrm{prob}(\{X_j\}|I). \qquad (7.2)$$

The term on the far right represents our state of knowledge about the $\{X_j\}$ given only the prior information I, while the new insight brought by the data is encoded in the likelihood $\mathrm{prob}(\{D_k\}|\{X_j\}, I)$. If the likelihood function is much more sharply peaked than the prior (with both being viewed in the space of the $\{X_j\}$), then it will dominate the posterior and we will have learnt a great deal from the (new) experimental results; if it is very broad, then the extra measurements add little to our existing position. Thus, if we have a choice in such matters, the experimental set-up should be tailored to make the likelihood function as sharply peaked as possible. In other words, the essence of optimal design is to make the data most sensitive to the quantities of interest.

While the general criterion above is very easy to state, and seems eminently reasonable, difficulties soon emerge when we consider its implementation in detail. For example, a proposed improvement in the conduct of an experiment may lead the multivariate likelihood function to become narrower in some directions but wider in others. An assessment of its potential benefits is then not so straightforward, and requires a careful consideration of the relative importance of the accuracy with which the various parameters are inferred. Another problem, which we have already met in the coin-flipping case, is that the spread of the pdf $\mathrm{prob}(\{D_k\}|\{X_j\}, I)$ may depend on the actual (true) values of the X_j's; since these are not known beforehand, but are to be estimated, the planning of an optimized experiment can be rather awkward. In particular, it could mean that an instrument designed to minimize the time taken to confirm that the X_j's were close to some anticipated values may not be the best one for dealing with the situation if our expectations turn out to be incorrect. Despite the many difficulties, however, it's still well worth thinking about the question of optimal experimental design: even if we can only manage a rather crude probabilistic analysis, the insight it provides can help in guiding us in the right direction.

7.2 Example 7: optimizing resolution functions

To illustrate some of the issues raised in the preceding section, let's consider the case of a simple convolution problem. We've used this example earlier, in Figs 4.2 and 6.10, where the data were related to the object of interest through a blurring process. While this smearing is a generally undesirable feature, it is often quite difficult to avoid in practice. Nevertheless, it may be possible to change the shape and width of the resolution, or point-spread, function to some extent by altering the design of the experimental set-up; how can this be done to our best advantage?

To make the analysis easier, let's assume that we are dealing with a one-dimensional problem (defined by the parameter x); let's also suppose that the nature of the blurring does not depend on location of the structure, so that it is described by an invariant resolution function $R(x)$. In many situations, this point-spread function is (very nearly) Gaussian in shape:

$$R(x) \propto \exp\left[-\frac{x^2}{2\,w^2}\right],\tag{7.3}$$

but it can be highly asymmetric, like a sharp-edged exponential:

$$R(x) \propto e^{-x/\tau} \quad \text{for } x \geqslant 0,\tag{7.4}$$

and zero otherwise. Although the truncated form of eqn (7.4) may seem extreme, variants of it are typical when x is time and R is a causal response: Charter (1990) gives an example from drug absorption studies, and Fig. 6.10(b) resembles the resolution function for experiments at pulsed neutron sources. Whatever the shape of $R(x)$, the 'ideal' data $\{F_k\}$ are related to the spectrum of interest $f(x)$ by the integral equation

$$F_k = T \int f(x)\,R(x_k - x)\,\mathrm{d}x + B(x_k),\tag{7.5}$$

for $k = 1, 2, \ldots, N$, where we have added the term $B(x)$ to allow for a background signal. The parameter T is a scaling constant whose magnitude may be proportional to the amount of time for which the experiment is conducted, but is independent of the width of the point-spread function (e.g. w or τ); the size of the background signal may also increase linearly with the recording time, but we will assume that this variation is already incorporated in $B(x)$.

In Section 7.1, we noted that the impact of the new data $\{D_k\}$ becomes greater as the corresponding likelihood function becomes narrower; as such, to optimize the set-up, we must investigate how the spread of $\mathrm{prob}(\{D_k\} \mid f(x), T, R(x), B(x), I)$ depends on the experimental parameters. If the measurements involve a counting process (e.g. the detection of neutrons or photons) then, as we saw in Chapter 5, the appropriate likelihood function is typically a Poisson pdf:

$$\mathrm{prob}(\{D_k\} \mid \{F_k\}, I) = \prod_{k=1}^{N} \frac{F_k^{D_k}\,e^{-F_k}}{D_k!},\tag{7.6}$$

where, for brevity, we have replaced $f(x)$, T, $R(x)$ and $B(x)$ in the conditioning statement by $\{F_k\}$, which are related to the former through eqn (7.5). When we talk about looking at the width of the likelihood function, we really mean that we want to know what range of different $f(x)$'s will give reasonable agreement with the data; to quantify this, we must specify how $f(x)$ is to be described. Hence, it's not enough merely to write down an expression such as eqn (7.6): as usual, we also have to think carefully about what we know about the nature of the object of interest. Consequently, the optimal strategy for one situation may not be the best one for another. Nevertheless, let's consider a couple of specific cases and see what we can learn.

7.2.1   An isolated sharp peak

Perhaps the simplest example that we can investigate concerns the amplitude and position of a sharp, isolated, peak. That is to say, we believe that the spectrum of interest consists of a single δ-function:

$$f(x) \;=\; A\,\delta(x-\mu)\,, \tag{7.7}$$

so that $f(x)$ is defined completely by the two parameters A and μ. With this form, the convolution integral of eqn (7.5) reduces to

$$F_k \;=\; A\,T\,R(x_k-\mu) + B(x_k)\,. \tag{7.8}$$

Returning to our central task of assessing how the parameters of the experimental design affect the width of the likelihood function, the easiest way of proceeding is to use a numerical brute force and ignorance method: (i) create a set of 'mock' data $\{D_k\}$, computer-generated to conform with the model of eqn (7.8) and the Poisson process of eqn (7.6); (ii) evaluate the probability in eqn (7.6), resulting from these simulated measurements, for different possible values of A and μ; (iii) display this two-dimensional pdf in (A, μ)-space as a contour map, to give a vivid representation of the spread of the likelihood function; (iv) examine how the range of A and μ values which yield reasonable agreement with the data changes as the experimental parameters (such as w or τ or T etc.) are varied.

The algorithm outlined above is actually akin to the procedure used to obtain the results in Fig. 3.3, where we analysed simulated Poisson data (for a signal peak on a flat background) generated under various 'experimental' conditions. This led to a brief discussion of how factors such as the number of counts, the spatial extent of the measurements, and so on, influenced the reliability with which the parameters of interest could be inferred. Although our observations were (naturally) based on the changes in the width and orientation of the resultant posterior pdfs, our conclusions would have been much the same had we considered the likelihood function instead; this is simply because these pdfs are proportional to each other, with the assignment of a naïve uniform prior (give or take the positivity cut-offs). An interesting feature of this approach to tackling design issues is that we don't need to create, and analyse, a large number of 'mock' data-sets for each set-up: just one will suffice! The reason is that while different realizations of random noise shift the maximum of the likelihood function, its spread remains largely unchanged; since our concern here is purely with the latter, we need only carry out one simulation for each situation.

Let's begin by considering the case of a Gaussian resolution function, as in eqn (7.3); if the background signal is also taken to be constant, so that $B(x) = B$, then the problem becomes very similar to that in Section 3.1. Rather than knowing μ (and w) and wanting to infer A and B, however, we will now suppose that B is given and it is A and μ that are to be estimated. The left-hand side of Fig. 7.1 shows four sets of data which have been simulated to investigate the effect of varying the design parameters T, w and B. The corresponding likelihood functions on the right-hand side are plotted with contour-levels at 10, 30, 50, 70 and 90% of the maximum probability. The uppermost panels represent a standard, or reference, against which other changes are to be assessed; in arbitrary units, it has $T = 20$, $w = 16$ and $B = T/10$. In the second pair of plots, the counting time has been increased ten-fold (so that $T = 200$, but w and B are still 16 and $T/10$); as might have been expected, the likelihood function becomes about three (or $\sqrt{10}$) times narrower in each direction. The next two panels show that a more curious result is obtained when the width of the resolution function is reduced to $w = 4$ (but T and B remain unchanged): the likelihood function shrinks by a factor of two with respect to μ, but expands to double its original size in the A-direction. This can be understood qualitatively by noting that a narrower $R(x)$ reduces both the amount of blurring and the total number of counts measured; while the former leads to an improvement in the positional discrimination provided by the data, the latter has an adverse effect on our ability to infer the amplitude reliably. Thus, despite the overt simplifications of our example, we are again faced with the realization that addressing experimental design issues is not a straightforward problem: what do we really mean by optimal, when we are forced to trade-off the accuracy of one parameter against another? The answer to this question can only be found by deeper soul-searching about our ultimate goal, and thence an assessment of the relative importance of the reliability of the parameters; although this awkwardness can be unsettling, the likelihood analysis serves a useful purpose in pointing us to matters which need more thought. Finally, the last two plots in Fig. 7.1 show how an increase in the level of the background signal (to $B = T/2$) degrades the quality of the data; this broadening of the likelihood function might be even more pronounced if the value of B were not assumed to be known, and had to be marginalized out of the problem.

Having seen how the effects of changing T, w and B can be investigated by the use of a few computer simulations, and a numerical (and pictorial) analysis of the likelihood function, let's try to understand these results from a more theoretical point of view. As in Fig. 7.1, suppose that the 'true' values are $A = 1$ and $\mu = 0$; then, according to eqns (7.3) and (7.8), the ideal data $\{\hat{D}_k\}$ will be given by

$$\hat{D}_k = T \exp\left[-\frac{x_k^2}{2\,w^2}\right] + B, \tag{7.9}$$

for $k = 1, 2, \ldots, N$. Although the measured counts $\{D_k\}$ will be related to these through a Poisson process, it's more convenient to approximate this corruption by an additive Gaussian noise term $\{\epsilon_k\}$:

$$D_k = \hat{D}_k + \epsilon_k, \tag{7.10}$$

where $\langle \epsilon_k \rangle = 0$ and $\langle \epsilon_k^2 \rangle = \sigma_k^2 = \hat{D}_k$. Within this context, the likelihood function of eqn (7.6) reduces to the form

Experimental design

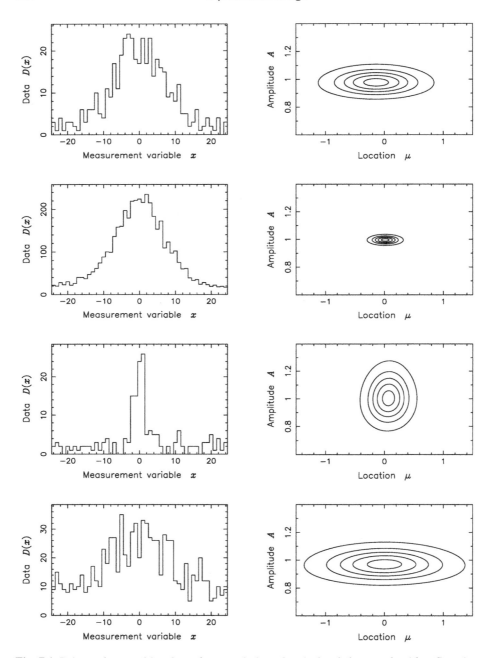

Fig. 7.1 Poisson data resulting from the convolution of an isolated sharp peak with a Gaussian resolution function, and a background signal, simulated under different experimental conditions. The corresponding likelihood functions are plotted with respect to the parameters to be inferred.

$$\log_e\left[\operatorname{prob}(\{D_k\}|\{F_k\}, I)\right] \approx \text{constant} - \chi^2/2, \qquad (7.11)$$

where χ^2 is the usual sum of the normalized-residual squareds,

$$\chi^2 = \sum_{k=1}^{N} \frac{(F_k - D_k)^2}{\sigma_k^2}, \qquad (7.12)$$

and the $\{F_k\}$ are related to the 'trial' values of A and μ by eqns (7.3) and (7.8):

$$F_k = A\, T \exp\left[-\frac{(x_k - \mu)^2}{2\,w^2}\right] + B. \qquad (7.13)$$

The maximum of $\operatorname{prob}(\{D_k\}|\{F_k\}, I)$ is, of course, defined by the condition that both $\partial\chi^2/\partial A$ and $\partial\chi^2/\partial\mu$ are equal to zero; its position will be fairly close to the true input-value, so that $A_O \approx 1$ and $\mu_O \approx 0$. Our interest here, however, is with the spread of the likelihood function about this optimal point; to study this, we must calculate the second partial derivatives $\partial^2\chi^2/\partial A^2$, $\partial^2\chi^2/\partial\mu^2$ and $\partial^2\chi^2/\partial A\,\partial\mu$.

Substituting for F_k and D_k from eqns (7.9), (7.10) and (7.13) in the formula for χ^2 of eqn (7.12), and then differentiating twice with respect to A while keeping μ constant, we obtain

$$\frac{\partial^2\chi^2}{\partial A^2} = \sum_{k=1}^{N} \frac{2\,T^2}{\sigma_k^2} \exp\left[-\frac{(x_k - \mu)^2}{w^2}\right]. \qquad (7.14)$$

This comprises one of the elements of the $\nabla\nabla\chi^2$ matrix, but it must be evaluated at the maximum of the likelihood function. Since we expect μ_O and A_O to be about zero and unity, respectively, this component will be approximately given by

$$\left.\frac{\partial^2\chi^2}{\partial A^2}\right|_{A_O, \mu_O} \approx \sum_{k=1}^{N} \frac{2\,T^2}{\sigma_k^2} \exp\left[-\frac{x_k^2}{w^2}\right]. \qquad (7.15)$$

After substituting $\sigma_k^2 = \hat{D}_k$ from eqn (7.9), the behaviour of the resulting summation can be ascertained by analogy with the integral

$$\int_{-\infty}^{\infty} \frac{\exp(-x^2/w^2)}{T\exp(-x^2/2w^2) + B}\, dx \approx \frac{w\sqrt{2\pi}}{T + B\sqrt{2}}, \qquad (7.16)$$

where the expression on the right-hand side is asymptotically correct in the limits of $B \to 0$ and $B \gg T$, and is accurate to within about 5% in the intermediate regime. This means that, as long as the data are measured finely enough over the entire region of the blurred peak, eqn (7.15) reduces to

$$\left.\frac{\partial^2\chi^2}{\partial A^2}\right|_{A_O, \mu_O} \propto \frac{w\,T^2}{T + B\sqrt{2}}. \qquad (7.17)$$

The other second partial derivatives of χ^2 can be obtained in a similar manner, but the corresponding algebra requires a little more effort and thought; to a reasonable approximation, it can be shown that

$$\left.\frac{\partial^2 \chi^2}{\partial \mu^2}\right|_{A_\mathrm{o},\mu_\mathrm{o}} \propto \frac{T^2}{w\left(T + B\sqrt{8}\right)} \qquad \text{and} \qquad \left.\frac{\partial^2 \chi^2}{\partial A\,\partial \mu}\right|_{A_\mathrm{o},\mu_\mathrm{o}} \approx 0. \qquad (7.18)$$

Since the covariance matrix is given by (twice) the inverse of $\nabla\nabla\chi^2$, we now have all the ingredients necessary for investigating how T, w and B influence the width, and orientation, of the likelihood function.

The first thing to notice is that the off-diagonal term $\partial^2 \chi^2/\partial A\,\partial \mu$ is negligible, as expected from left–right symmetry. Not only does this make it very easy to invert the $\nabla\nabla\chi^2$ matrix, it also tells us that the data impose no correlation between the amplitude and position of the sharp peak. This property is confirmed by Fig. 7.1, because the principal directions of the likelihood ellipses are parallel to the coordinate axes (rather than being skew). The covariance matrix for A and μ becomes particularly simple in the limit of a small background signal ($B \ll T$), when it takes the form

$$\begin{pmatrix} \langle \delta\mu^2 \rangle & \langle \delta\mu\,\delta A \rangle \\ \langle \delta\mu\,\delta A \rangle & \langle \delta A^2 \rangle \end{pmatrix} \propto \frac{1}{T}\begin{pmatrix} w & 0 \\ 0 & 1/w \end{pmatrix}, \qquad (7.19)$$

where $\delta\mu = \mu - \mu_\mathrm{o}$, and so on. Hence, as expected, the range of parameter values which give reasonable agreement with the data is inversely proportional to $\sqrt{T}$; in other words, a ten-fold increase in the counting time for the experiment causes the likelihood function to sharpen-up by a factor of $\sqrt{10}$. The peculiar behaviour of $\mathrm{prob}(\{D_k\}|\{F_k\}, I)$ with respect to changes in the width of the Gaussian resolution function can also be understood from eqn (7.19): $\langle \delta\mu^2 \rangle \propto w$, whereas $\langle \delta A^2 \rangle \propto 1/w$. As we said earlier, this opposite dependence of A and μ on w illustrates one of the dilemmas often faced when trying to optimize an experimental design: we can only improve the accuracy with which one of the parameters is determined at the cost of degrading the precision for another! Finally, eqns (7.17) and (7.18) also indicate how an increase in the background signal has an adverse effect on the quality of the data. A larger value of B reduces the magnitude of both $\partial^2 \chi^2/\partial A^2$ and $\partial^2 \chi^2/\partial \mu^2$, by essentially raising the level of the (Poisson) noise σ_k^2; the inverse of the $\nabla\nabla\chi^2$ matrix then gives the corresponding increase in the range of the A and μ values which become consistent with the measurements.

7.2.2   A free-form solution

An obvious way of extending the preceding analysis is to consider a spectrum having two sharp peaks. Not only does this open up the possibility of investigating how different experimental conditions affect the reliability with which we can infer the four relevant parameters (A_1, A_2, μ_1 and μ_2), it also enables a study of our ability to resolve two closely spaced features if we're not sure. Rather than pursuing this generalization, and then moving on to several peaks, or specific types of broad structure, let's proceed straight to the (extreme) situation where we know very little about the spectrum $f(x)$.

In other words, while the experimental set-up is still defined by eqns (7.5) and (7.6), we need to use the free-form representation of $f(x)$ discussed in Section 6.1:

$$f(x) = \sum_{j=1}^{M} a_j \, \delta\left(x - x_j\right),$$ (7.20)

where M is large and the $\{x_j\}$ are fixed by eqn (6.3). With this formulation, our inference about $f(x)$ is described by the posterior pdf for the M amplitude coefficients $\{a_j\}$; we must, therefore, view the likelihood function in the multidimensional space of these parameters, when examining how its spread depends on the experimental conditions.

The main difference between the free-form situation and the earlier case is that we are now dealing with a much larger multivariate problem; rather than just having the two parameters A and μ, there may be hundreds of a_j! Hence, in contrast to Fig. 7.1, it will be impossible to present a direct picture of the likelihood function. Nevertheless, we can still carry out a theoretical analysis within the quadratic approximation of eqns (3.30)–(3.32); in that context, L is now the (natural) logarithm of $\mathrm{prob}(\{D_k\}|\{F_k\}, I)$ and the $\{X_j\}$ are to be read as $\{a_j\}$. Thus, with the simplification afforded by eqn (7.11), the width and orientation of the likelihood function is characterized by the second-derivative matrix

$$\left[\boldsymbol{\nabla}\boldsymbol{\nabla}\chi^2\right]_{ij} = \frac{\partial^2 \chi^2}{\partial a_i \, \partial a_j}.$$ (7.21)

Substituting for $f(x)$ from eqn (7.20) in eqn (7.5), and then differentiating the resultant expression of eqn (7.12), we obtain:

$$\left[\boldsymbol{\nabla}\boldsymbol{\nabla}\chi^2\right]_{ij} = \sum_{k=1}^{N} \frac{2T^2}{\sigma_k^2} R(x_k - x_i) R(x_k - x_j).$$ (7.22)

The inverse of $\boldsymbol{\nabla}\boldsymbol{\nabla}\chi^2$ does, of course, yield the covariance matrix, which tells us the range of parameter values $\langle \delta a_j^2 \rangle$ which give reasonable agreement with the data; its off-diagonal terms give the related correlations $\langle \delta a_i \, \delta a_j \rangle$. Although this is the type of information we need for addressing questions of experimental design, it is much more convenient to ascertain this from the eigenvalues and eigenvectors of eqn (7.22). With reference to Fig. 3.6, and our singular value decomposition discussion in Section 6.1, this is simply because it's a lot easier to think about the spread of a skew likelihood function in terms of the widths along its principal axes.

As in eqn (6.6), the eigenvalues $\{\lambda_l\}$ and eigenvectors $\{e_l(x)\}$ of the $\boldsymbol{\nabla}\boldsymbol{\nabla}\chi^2$ matrix are given by the solutions of the standard equation:

$$\sum_{i=1}^{M} \left[\boldsymbol{\nabla}\boldsymbol{\nabla}\chi^2\right]_{ij} e_l(x_i) = \lambda_l \, e_l(x_j),$$ (7.23)

for $l = 1, 2, \ldots, M$. These can be obtained numerically, as in Figs 6.2 and 6.3, by simulating a set of noisy data from a test spectrum, computing the elements of the resulting $\nabla\nabla\chi^2$ matrix in eqn (7.22) and solving eqn (7.23). Alternatively, we can do the calculations analytically by making suitable simplifying approximations; as we saw a little earlier, these are often aided by considering analogies with the continuum limit. In this instance, for example, the summation of eqn (7.22) can be replaced by the integral

$$\nabla\nabla\chi^2(x, x') \;=\; \frac{2T^2}{\sigma^2} \int R(y-x)\, R(y-x')\,\mathrm{d}y\,, \tag{7.24}$$

as long as the data have been sampled fairly finely, and the size of the error-bars is roughly constant (so that $\sigma_k \approx \sigma$). This is, in fact, just the *auto-correlation* function of $R(x)$, whose magnitude depends only on the separation between the points x and x'; denoting this symmetric function by G, we have

$$\nabla\nabla\chi^2(x, x') \;=\; \frac{2T^2}{\sigma^2}\, G\big(|x-x'|\big)\,. \tag{7.25}$$

Substituting this into the continuum version of eqn (7.23), the eigenvalue equation then takes the form

$$\frac{2T^2}{\sigma^2} \int G\big(|x-x'|\big)\, e_l(x')\,\mathrm{d}x' \;=\; \lambda_l\, e_l(x)\,. \tag{7.26}$$

Since the left-hand side is a convolution integral, eqn (7.26) is best solved by taking its Fourier transform. This allows us to write

$$\frac{2T^2}{\sigma^2}\, \hat{G}(\omega)\, \hat{e}_l(\omega) \;=\; \lambda_l\, \hat{e}_l(\omega)\,, \tag{7.27}$$

where the real-space x-functions are related to their (complex) ω-hat Fourier counterparts by eqn (6.24). For any arbitrary $R(x)$, the only thing we can say about $\hat{G}(\omega)$ is that it must be real and symmetric; hence, by inspection, the most general solutions of eqn (7.27) have to be of the type

$$\hat{e}_l(\omega) \;\propto\; \delta(\omega-\omega_l) \pm \delta(\omega+\omega_l)\,, \tag{7.28}$$

where ω_l is any real number, the eigenvalues being proportional to $\hat{G}(\omega_l)$. The (inverse) Fourier transform of eqn (7.28) then tells us that the eigenfunctions for the convolution problem are

$$e_l(x) \;\propto\; \sin(\omega_l x) \quad\text{and}\quad \cos(\omega_l x)\,. \tag{7.29}$$

The corresponding eigenvalues are given by eqn (7.27) as:

$$\lambda_l \;=\; \frac{2T^2}{\sigma^2}\, |\hat{R}(\omega_l)|^2\,, \tag{7.30}$$

where we have used a standard result about the Fourier transform of auto-correlation functions to replace $\hat{G}(\omega)$ with $|\hat{R}(\omega)|^2$.

In Section 6.1, we noted that the eigenvectors of the $\nabla\nabla\chi^2$ matrix represented the linear combinations of the $\{a_j\}$ in eqn (7.20) which could be estimated independently of each other. Hence, if we express the spectrum of interest as a sum of the $\{e_l(x)\}$ in eqn (7.29):

$$f(x) = \sum_l \left[s_l \sin(\omega_l x) + c_l \cos(\omega_l x) \right], \qquad (7.31)$$

and view $\mathrm{prob}(\{D_k\}|f(x), I)$ in the (M-dimensional) space of the coefficients $\{s_l\}$ and $\{c_l\}$, then the principal directions of the (elliptical) likelihood function will be parallel to these coordinate axes; the corresponding widths will be given by the reciprocal of the eigenvalues in eqn (7.30):

$$\langle \delta s_l^2 \rangle = \langle \delta c_l^2 \rangle = 2/\lambda_l. \qquad (7.32)$$

Since the larger values of ω_l give rise to more rapidly varying structure in $f(x)$, these frequencies are inversely related to the fineness of the detail in the spectrum of interest (so that $\delta x \propto 1/\omega$). Therefore, the error-bars for the various sine and cosine contributions in eqn (7.32) tell us how much information the data contain about features of $f(x)$ on different length-scales.

To explain the last point further, consider the two resolution functions shown in Fig. 7.2(a). They are just the Gaussian and sharp-edged exponential of eqns (7.3) and (7.4), with w and τ chosen to yield the same full-width at half-maximum; their peak heights have also been adjusted to give them equal integrated areas. According to eqn (7.30), the resulting eigenvalue spectra $\sqrt{\lambda(\omega)}$ are proportional to the moduli of the Fourier transforms of these point-spread functions $|\hat{R}(\omega)|$; they are plotted in Figs 7.2(b) and (c). The highest frequency, $\omega = 64$, corresponds to pixel-scale structure $\delta x \approx 1$, while the lowest, $\omega = 0$, refers to features in $f(x)$ which do not vary at all with position. Thus the fall-off in the eigenvalue spectra, with increasing ω, confirms our expectation that information about the finer aspects of $f(x)$ is lost in a blurring process. In fact, judging by the different decay-rates of the $\lambda(\omega)$, Fig. 7.2(b) tells us that this degradation is far more severe for the Gaussian $R(x)$ than it is for the exponential one; this can be seen clearly from the logarithmic plot of Fig. 7.2(c). We should point out, however, that this sharp contrast arises primarily from the presence of the hard edge in the exponential $R(x)$, and not from the $\exp(-x)$ and $\exp(-x^2)$ decays.

The profound effect that the nature of a resolution function can have on the reliability with which intricate detail can be inferred is illustrated in Fig. 7.3. It shows the best estimates of $f(x)$ that result from a MaxEnt analysis of two simulated data-sets, which were generated by convolving a test spectrum with the functions of Fig. 7.2(a); as we would have anticipated, the sharp features are (faithfully) recovered in the exponential case but are lost with the Gaussian blur.

Before finally leaving this example, let's see how changes in the background level, the experimental time and the width of the resolution function affect the quality of the data. To do this, we need to consider the noise term σ^2 in eqn (7.30) a little more carefully. This quantity arose from making the simplifying approximation that all the $\{\sigma_k\}$ in eqn (7.12) were roughly equal; for Poisson data, its magnitude will be proportional to the total number of counts measured. Thus, we would expect that

$$\sigma^2 \propto T R_0 \left(1 + B_f\right), \qquad (7.33)$$

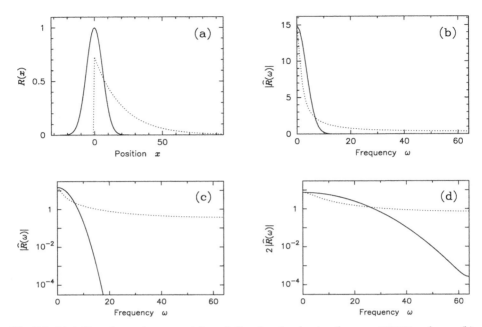

Fig. 7.2 (a) A Gaussian and exponential resolution function having the same FWHM and area. (b) The modulus of their Fourier transforms. (c) The same as (b), but on a logarithmic scale. (d) The equivalent of (c), but for a Gaussian and exponential having only a quarter of the width in (a).

where B_f is the fraction of counts emanating from the background signal, as compared to those from the spectrum of interest, and R_0 is the integral of $R(x)$. Substituting this into eqn (7.30), we obtain the familiar result that $\lambda \propto T$; in other words, the likelihood function sharpens up with the square root of the experimental time. The way in which a large background level degrades the quality of the data, by contributing to the noise σ, is also apparent from eqn (7.33). The strong influence that the shape of the resolution function can have on the convolution problem has already been noted. This property is formally encoded in the characteristics of the eigenvalue spectrum, with eqn (7.30) showing that the latter is primarily governed by the Fourier transform of $R(x)$; according to eqn (7.33), there is also a weak dependence on the integrated area R_0. The effect of changing the width of the point-spread function is illustrated in Fig. 7.2(d), which gives $|\tilde{R}(\omega)|/\sqrt{R_0}$ corresponding to w and τ having a quarter of the values in Fig. 7.2(a). We see that while this four-fold narrowing of the resolution functions makes an enormous difference to the $\lambda(\omega)$ for the Gaussian, there is little change in the exponential case. This acute sensitivity of the eigenvalue spectrum on w tells us that the loss in resolving-power incurred from a broad Gaussian blur cannot easily be compensated for by an increase in the measurement time; for that, we would have to boost T by many orders of magnitude! This behaviour contrasts markedly with that of the exponential, which seems to preserve fine detail information largely irrespective of τ, and indicates there is great (potential) merit in having sharp features in the point-spread function.

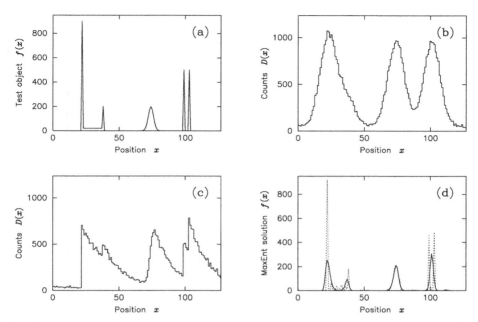

Fig. 7.3 (a) A test spectrum, on a grid of 128 pixels. (b) Simulated (noisy) data generated by convolving (a) with the Gaussian in Fig. 7.2(a). (c) The same as (b), but using the exponential in Fig. 7.2(a). (d) The (free-form) MaxEnt estimates of $f(x)$ resulting from the data in (b) and (c), plotted as a solid and dotted line respectively.

7.3 Calibration, model selection and binning

One of the problems which frequently plagues experimental studies is that of nuisance parameters. Although we know how to deal with these quantities formally, through marginalization, a relevant design issue is how much effort should be spent in trying to calibrate them out. To take a specific example, consider the case of the (flat) background signal in the preceding analysis; if the value of B was not known, what fraction of our time should we devote towards its measurement? Well, given that we're interested in the parameters $\{X_j\}$ and not B, we really need to look at the spread of the marginal likelihood function $\mathrm{prob}(\{D_k\}|\{X_j\}, I)$ rather than the conditional pdf $\mathrm{prob}(\{D_k\}|\{X_j\}, B, I)$; the two are related by

$$\mathrm{prob}(\{D_k\}|\{X_j\}, I) = \int \mathrm{prob}(\{D_k\}|\{X_j\}, B, I)\,\mathrm{prob}(B|I)\,\mathrm{d}B, \qquad (7.34)$$

where we have dropped the conditioning of B on the $\{X_j\}$ in the last term, because we are concerned with the merit of obtaining an independent estimate of the background. If the data $\{D_k\}$ are collected for a time T_D, then the width of $\mathrm{prob}(\{D_k\}|\{X_j\}, B, I)$ will tend to be proportional to $1/\sqrt{T_D}$; similarly, if a time T_B is spent on a background calibration, then the spread of $\mathrm{prob}(B|I)$ will scale like $1/\sqrt{T_B}$. Since both these pdfs contribute to the precision attainable for the parameters $\{X_j\}$, the best choice of the ratio $T_B : T_D$ is simply the one that makes the integral of eqn (7.34) as sharply peaked as possible. While the optimal value of this temporal division will depend on the details of the problem at hand, a simple illustration of what we are trying to do was provided by Fig. 3.4. In that example, the $\{X_j\}$ pertain to just one quantity; namely $X_1 = A$, the amplitude of the isolated peak. Figure 3.4 shows the comparison between the marginal pdf for A, where $\mathrm{prob}(B|I) = \mathrm{constant}$ (for $B > 0$), with the one conditional on knowing the true value of B; this is akin to considering the extreme cases of $T_B = 0$ and $T_B \to \infty$, respectively, for a fixed T_D. Judging by the similarity of the widths of these pdfs, we see that a separate measurement of the background signal is of limited benefit in three out of the four situations discussed. Our conclusions would be very different, however, if $B(x)$ was highly structured (rather than slowly varying); in such circumstances, we have little choice but to spend an appreciable amount of time on a calibration experiment.

So far, we have concentrated on the optimization of experimental design for parameter estimation problems. In other words, we've assumed a definitive model H_1 which allows us to characterize the object of interest by a specific set of variables $\{X_j\}$; although their number can be rather large for a free-form solution, our central strategy has been to make the likelihood function as sharply peaked as possible in order to maximize the impact of the (new) data. What should we do, however, if our primary objective is to choose between H_1 and an alternative hypothesis H_2? Well, we are then concerned with the ratio of their posterior probabilities

$$\frac{\mathrm{prob}(H_1|\{D_k\},I)}{\mathrm{prob}(H_2|\{D_k\},I)} = \frac{\mathrm{prob}(H_1|I)}{\mathrm{prob}(H_2|I)} \times \frac{\mathrm{prob}(\{D_k\}|H_1,I)}{\mathrm{prob}(\{D_k\}|H_2,I)}, \qquad (7.35)$$

where we have used Bayes' theorem, top and bottom, to obtain the right-hand side. As the ratio of the priors for H_1 and H_2 is likely to be of order unity, a well-designed experiment will be one that makes the contrast between their evidences as large as possible. Using marginalization and the product rule, and assigning a simple uniform prior for their respective model parameters, the latter are given by

$$\mathrm{prob}(\{D_k\}|H_1,I) = \int\!\!\int \cdots \int \mathrm{prob}(\{D_k\}|\{X_j\},H_1,I)\,dX_1\,dX_2\cdots dX_M, \quad (7.36)$$

and a similar expression with H_2 and Y_j's. Since the evidence is proportional to the integral of the likelihood function, we need to consider its height as well as its spread; in terms of eqn (7.11), for example, this means that both the minimum value of χ^2 and the determinant of $\nabla\nabla\chi^2$ are important. In fact, as the goodness-of-fit term tends to be dominant, a model selection experiment will be optimized if most of the data are collected where the competing hypotheses give (drastically) different predictions. Having said that, there might still be merit in the regions of commonality if we have an intrinsic interest in the parameters of the favoured model; this will be particularly so if it greatly sharpens up the likelihood function.

The discussion above serves as a useful reminder of how the nature of the data-collection procedure can have a strong influence on what we can glean from an experiment. That is to say, it's not just the physical characteristics of the apparatus that are important but so too are the questions of where, and for how long, different measurements are taken. In Fig. 3.3, for example, we saw how changes in the x-range of the data could affect the precision with which the amplitude of a peak was inferred: while a very narrow window was bad because it led to a strong correlation with the background, there was also little point in going too far beyond the tails of the signal. Similarly, a fine sampling-rate can be important for a sharp-edged resolution function but is not required for a smooth one. This can be understood analytically by considering the impact that a bin-width Δx will have on the eigenvalue spectrum of Fig. 7.2. A discrete Fourier transform of $R(x)$, digitized on this length-scale, will yield an upper bound on the frequency of $\omega_{max} \sim 1/\Delta x$; this will, in turn, place a limit on the fineness of the detail in $f(x)$ that can be inferred reliably, unless $\lambda(\omega)$ has already decayed to almost zero by this cut-off. Hence, a high data sampling-rate is far more crucial for a point-spread function with sharp features than it is for a broad smooth one.

7.4   Information gain: quantifying the worth of an experiment

In the general discussion of Section 7.1, it was noted that a good experiment was one that yielded a likelihood function which was much more sharply peaked than the prior; otherwise, very little is learnt from the measurements. This basic idea can be quantified through the notion of entropy that was introduced in Chapter 5, eqns (5.26)–(5.29). Here, we reverse the sign, and define

$$\mathcal{H} = \int P(X) \log_2 \left[\frac{P(X)}{\pi(X)} \right] dX \, . \tag{7.37}$$

This negative entropy is the (positive) amount of *information* in the posterior

$$P(X) = \text{prob}(X|D, I)$$

relative to the prior

$$\pi(X) = \text{prob}(X|I) \, , \tag{7.38}$$

after acquiring data D. With logarithms to the base 2, information is measured in *bits*. If X had just two equivalent states, for example, then the prior would be $\pi(X) = (\frac{1}{2}, \frac{1}{2})$. Full knowledge of X, on the other hand, would compress its posterior $P(X)$ to either $(1, 0)$ or $(0, 1)$. Either corresponds to an information of

$$\mathcal{H} = 1 \log_2(2) + 0 \log_2(0) = 1 \text{ bit} \, ,$$

as $P \log P \to 0$ for small P. Similarly, with four equivalent states and prior $\pi(X) = (\frac{1}{4}, \frac{1}{4}, \frac{1}{4}, \frac{1}{4})$, full knowledge that compresses the posterior into just one state has information $\mathcal{H} = 2$ bits. Knowing how a 6-face die landed gives $\mathcal{H} = \log_2 6 = 2.58$ bits, and so on. Equation (7.37) generalizes such examples to partial knowledge over an arbitrary prior. It is the corner-stone of *information theory*, as founded by Shannon (1948).

Now suppose that we design an experiment to measure X (either a single scalar or a vector of several parameters). The experiment is to produce data D (again, either scalar or vector), and is specified by its likelihood function

$$\mathcal{L}(D, X) = \mathrm{prob}(D \,|\, X, I)\,.$$

The joint distribution $\mathrm{prob}(D, X \,|\, I) = \mathcal{L}(D, X)\,\pi(X)$ yields, as usual, the evidence

$$Z(D) = \mathrm{prob}(D \,|\, I) = \int \mathcal{L}(D, X)\,\pi(X)\,\mathrm{d}X \qquad (7.39)$$

and the posterior

$$P(X) = \frac{\mathcal{L}(D, X)\,\pi(X)}{Z(D)}\,. \qquad (7.40)$$

Here, though, the formulae remain general: we are only designing the equipment and have not yet acquired specific data. Nevertheless, we can use eqn (7.37) to discover the amount of information

$$\mathcal{H}(D) = \int \frac{\mathcal{L}(D, X)\,\pi(X)}{Z(D)}\,\log_2\!\left[\frac{\mathcal{L}(D, X)}{Z(D)}\right]\mathrm{d}X$$

that we would have about X if we did the experiment. Even before acquiring the data, we can anticipate the results with eqn (7.39), which gives us the expected information

$$\langle \mathcal{H} \rangle = \int \mathcal{H}(D)\,Z(D)\,\mathrm{d}D\,.$$

This is the *benefit* of the experiment, quantifying the amount of information about X which the experiment is expected to provide.

Sometimes we are not interested in X directly, but only in some subsidiary property $Q(X)$, perhaps with nuisance parameters eliminated. In that case, we use the prior and posterior of X in eqns (7.38) and (7.40) to induce prior and posterior for Q instead, through

$$\mathrm{prob}(Q^* \,|\, \ldots) = \int \delta\big(Q^* - Q(X)\big)\,\mathrm{prob}(X \,|\, \ldots)\,\mathrm{d}X\,.$$

We can then obtain the benefit of the proposed experiment in terms of expected information about Q instead of X.

The subject of 'Bayesian design' is the study of such expectations, and of the trade-offs that have to be made between benefits and costs. It puts experimental design on a firm, quantitative footing.

8 Least-squares extensions

Least-squares is probably the most widely used data analysis procedures in the physical sciences, and we saw how it could be justified from a Bayesian viewpoint in Chapter 3. The deeper understanding provided by the latter leads naturally to powerful extensions of the basic recipe. A consideration of the uncertainties in the characteristics of the implicit Gaussian likelihood, for example, yields simple prescriptions for dealing with 'outliers' and suspected noise correlations.

8.1 Introduction: constraints and restraints

The least-squares procedure for estimating the M parameters X of a suitable model from a pertinent set of N data D is given by the minimum of the χ^2 function:

$$\chi^2 = \sum_{k=1}^{N} R_k^2 \,, \tag{8.1}$$

where $R_k = (F_k - D_k)/\sigma_k$ is the normalized residual for the kth datum, being the difference between the theoretical prediction F_k,

$$F_k = f(X, k) \,, \tag{8.2}$$

and the corresponding datum D_k, relative to an estimate of the expected mismatch σ_k. As we saw in Section 3.5, this prescription follows from the assignment of a uniform prior for the X,

$$\mathrm{prob}(X | I) = \mathrm{constant} \,, \tag{8.3}$$

and an uncorrelated Gaussian likelihood function for the D,

$$\mathrm{prob}(D | X, I) = \prod_{k=1}^{N} \frac{1}{\sigma_k \sqrt{2\pi}} \exp\left(-\frac{R_k^2}{2}\right) \propto \exp\left(-\frac{\chi^2}{2}\right), \tag{8.4}$$

so that the logarithm of the posterior pdf becomes

$$L = \log_e\left[\mathrm{prob}(X | D, I)\right] = \mathrm{constant} - \frac{\chi^2}{2} \,, \tag{8.5}$$

where a knowledge of the $\{\sigma_k\}$, and the functional relationship f, is assumed in the background information I. The least-squares procedure is fairly simple to apply and often quite successful in practice. The reasons for this are three-fold: (i) if $N \gg M$, the posterior pdf is usually dominated by the likelihood and the prior is largely irrelevant;

(ii) the Gaussian pdf is a good approximation in many situations (e.g. for a Poisson pdf where $D_k \gg 1$); (iii) the optimization task is particularly easy if the functional relationship in eqn (8.2) is linear.

With this basic understanding of the assumptions that underlie the use of least-squares, a straightforward extension is immediately obvious if there is cogent prior information about the model parameters. If we already knew that $X_j = x_{oj} \pm \epsilon_j$, for example, where $j = 1, 2, \ldots, M$, then the assignment of an uncorrelated Gaussian pdf for the prior for X,

$$\mathrm{prob}(X \,|\, I) = \prod_{j=1}^{M} \frac{1}{\epsilon_j \sqrt{2\pi}} \, \exp\left[-\frac{(X_j - x_{oj})^2}{2\,\epsilon_j^2} \right] \propto \exp\left(-\frac{C}{2} \right), \qquad (8.6)$$

where

$$C = \sum_{j=1}^{M} \left(\frac{X_j - x_{oj}}{\epsilon_j} \right)^2, \qquad (8.7)$$

leads to the following for the logarithm of the posterior pdf:

$$L = \log_e\left[\mathrm{prob}(X \,|\, D, I) \right] = \mathrm{constant} - \tfrac{1}{2}\left[\chi^2 + C \right]. \qquad (8.8)$$

Hence, the best estimate of the X is given by those values which minimize $\chi^2 + C$. This additional constraint function C is negligible if the initial uncertainties $\{\epsilon_j\}$ are very large, when the original least-squares procedure is recovered. The extra term becomes important, however, if the measurements are insensitive to the values of some of the X_j's, or if they only tell us about a restricted linear combination of them. In the limit of very strong prior information, when some of the $\epsilon_j \to 0$, the constraints are called *restraints*. It's equivalent to holding the relevant X_j's at their presumed values, with a corresponding reduction in the number of model parameters M.

The specification of a non-uniform prior is widely seen as the principal difference between the Bayesian and the orthodox approaches to data analysis. While this is significant in itself, because the addition of the C term to χ^2 above seems more natural than an *ad hoc* fix, the advantages over a conventional viewpoint are much broader. Even with the stated prior and likelihood assumptions inherent in least-squares, useful extensions still follow from a careful consideration of the underlying uncertainties.

8.2   Noise scaling: a simple global adjustment

Suppose we were faced with our generic parameter estimation problem, and were prepared to make the assignments of eqns (8.3) and (8.4), but had no estimates for the $\{\sigma_k\}$. How could we then proceed? Well, we'd have to make some suitable assumptions; the simpler the better. For example, that all the $\{\sigma_k\}$ were roughly the same and equal to the unknown constant σ; or that they were proportional to the square root of the data (Poisson-like), $\sigma_k = \sigma \sqrt{D_k}$; or whatever seemed most appropriate. This is quite routine from our viewpoint, where probabilities represent states of knowledge, because all analyses are conditional on the background information and assumptions that go into them. If there was serious doubt about the validity of a proposed noise assignment, then it could always be compared quantitatively with an alternative suggestion through a comparison of their resultant probabilistic evidence, $\text{prob}(data\,|\,assumptions, I)$, as in Chapter 4.

Pursuing the most elementary hypothesis, defined by just one extra parameter σ, the conditional likelihood for the data is still of the form of eqn (8.4):

$$\text{prob}(\boldsymbol{D}\,|\,\sigma, \boldsymbol{X}, I) \,\propto\, \frac{1}{\sigma^N} \, \exp\!\left(-\frac{\chi^2}{2\,\sigma^2}\right), \tag{8.9}$$

where all occurrences of σ have been highlighted explicitly. Having separated out the noise scaling factor in this way, χ^2 is simply given by eqn (8.1) with $\sigma_k = 1$ if we have no clues; $\sigma_k = \sqrt{D_k}$ for Poisson-like situations; $\{\sigma_k\}$ are equal to the supplied estimates if they're thought to be reasonable to within an overall multiplicative factor; and so on. The marginal likelihood, with σ integrated out with a Jeffreys' prior, $\text{prob}(\sigma\,|\,\boldsymbol{X}, I) = \text{prob}(\sigma\,|\,I) \propto 1/\sigma$, is then given by

$$\text{prob}(\boldsymbol{D}\,|\,\boldsymbol{X}, I) \;=\; \int\limits_0^\infty \text{prob}(\boldsymbol{D}\,|\,\sigma, \boldsymbol{X}, I)\,\text{prob}(\sigma\,|\,I)\,\text{d}\sigma \;\propto\; \int\limits_0^\infty \left(\frac{2t}{\chi^2}\right)^{\!\frac{N}{2}-1}\! \text{e}^{-t}\,\frac{\text{d}t}{\chi^2}\,,$$

where we have substituted $t = \chi^2/2\sigma^2$. The expression on the right is simply a definite integral of t, which is just a 'number', times $N/2$ factors of χ^2 in the denominator. With the uniform prior of eqn (8.3), therefore, the logarithm of the posterior pdf reduces to

$$L \;=\; \log_e\!\left[\,\text{prob}(\boldsymbol{X}\,|\,\boldsymbol{D}, I)\,\right] \;=\; \text{constant} \,-\, \frac{N}{2}\,\log_e(\chi^2)\,. \tag{8.10}$$

In fact, we have already met this result in Chapter 3: it's equivalent to eqn (3.38) and the penultimate equation in Section 3.5.1, to within a Jeffreys' prior, and to eqn (3.44). While the best estimate of the model parameters, $\boldsymbol{X}_O$, is still given by the $\boldsymbol{X}$ which yield the smallest value of χ^2, $\chi^2_{\min}$, the quadratic measure of their covariance, $(\boldsymbol{\nabla}\boldsymbol{\nabla}L)^{-1}$ evaluated at $\boldsymbol{X}_O$, becomes

$$\left\langle (X_i - X_{Oi})(X_j - X_{Oj}) \right\rangle \;=\; 2\left[(\boldsymbol{\nabla}\boldsymbol{\nabla}\chi^2)^{-1}\right]_{ij} \frac{\chi^2_{\min}}{N}\,, \tag{8.11}$$

instead of just twice the inverse of $\boldsymbol{\nabla}\boldsymbol{\nabla}\chi^2$ (at $\boldsymbol{X}_O$), where i and $j = 1, 2, \ldots, M$.

8.3 Outliers: dealing with erratic data

The fitting of a linear model to a pertinent set of data, such as those in Fig. 8.1(a), is probably the most frequent use of least-squares; the relevant algebra is in Section 3.5.1. Sometimes this procedure is used 'blindly', and Fig. 8.1(b) shows an example of what can happen if we are not careful. This illustration is so simple that it's easy to understand and rectify the problem: a few of the data do not conform to the straight line hypothesis, to within their estimated error-bars, and their presence distorts the analysis

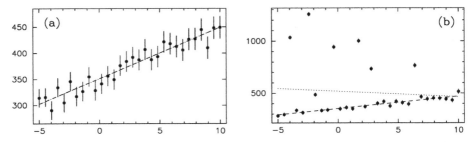

Fig. 8.1 The problem of outliers: (a) a 'well-behaved' set of data; (b) a case where quirky things occasionally happen. The least-squares estimate of the best straight lines is indicated by the dots, whereas the corresponding results following the analysis in Section 8.3.1 is marked by the dashes.

badly; assuming that the fault lies in the quirkiness of the measurements, rather than in the inadequacy of the model, the suspect data can be expunged. When the number of measurements is very large, or they're multidimensional in nature, or the fitting function is complicated, such a manual remedy may not be feasible. So, how can we deal with 'outliers' in an automatic fashion?

8.3.1 A conservative formulation

We have noted that the source of the difficulty is the presence of data which do not conform to our presumed model to within the stated error-bars. Although we could try to identify them, in some iterative procedure, let's treat all the measurements on an equal footing and adopt a more suspicious attitude towards all of them. That is to say, we'll treat the given $\{\sigma_k\}$ as merely representing lower bounds on the noise; after all, they are typically estimated under ideal conditions that sometimes fail to hold.

Consider a single datum, D. If all we know is that the expected mismatch with the model prediction F is greater than or equal to σ_o, what should we assign for the pdf for the error-bar σ? Since σ is a scale parameter, we might think of a Jeffreys' pdf:

$$\mathrm{prob}(\sigma|\sigma_\mathrm{o},\sigma_1,I) = \frac{1}{\log_e(\sigma_1/\sigma_\mathrm{o})} \times \frac{1}{\sigma} , \qquad (8.12)$$

for $\sigma_\mathrm{o} \leqslant \sigma < \sigma_1$, and zero otherwise. As this requires the specification of a finite upper bound, σ_1, let's use a variant that avoids this:

$$\mathrm{prob}(\sigma|\sigma_\mathrm{o},I) = \frac{\sigma_\mathrm{o}}{\sigma^2} , \qquad (8.13)$$

for $\sigma \geqslant \sigma_\mathrm{o}$, and zero otherwise; we will indicate the difference that this modification makes later. The marginal likelihood for D, with the unknown σ integrated out, is given by

$$\mathrm{prob}(D|F,\sigma_\mathrm{o},I) = \int\limits_0^\infty \mathrm{prob}(D|F,\sigma,I)\,\mathrm{prob}(\sigma|\sigma_\mathrm{o},I)\,\mathrm{d}\sigma , \qquad (8.14)$$

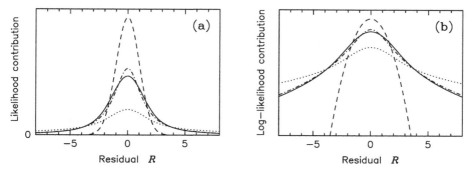

Fig. 8.2 (a) The likelihood contribution from a datum, $\mathrm{prob}(D|F, \sigma_\mathrm{o}, I)$, plotted as a function of the residual $R = (F-D)/\sigma_\mathrm{o}$, when σ_o is believed (dashed) compared with when it's used as a lower bound (solid); a Jeffreys' prior solution for the latter case, with $\sigma_1 = 100\,\sigma_\mathrm{o}$, is also shown (dotted). The dot–dashed line is the Cauchy variant of the Gaussian (dashed), as described in Section 8.3.3. (b) The same curves plotted on a logarithmic axis.

where we have dropped the conditioning on σ_o in the first term on the right-hand side, and F in the second, as being unnecessary. Making the least-squares assumption of a Gaussian pdf for $\mathrm{prob}(D|F, \sigma, I)$, and using the assignment of eqn (8.13), we obtain

$$\mathrm{prob}(D|F, \sigma_\mathrm{o}, I) \;=\; \frac{\sigma_\mathrm{o}}{\sqrt{2\pi}} \int_0^{1/\sigma_\mathrm{o}} t\, e^{-t^2(F-D)^2/2}\, \mathrm{d}t\,,$$

where we have substituted $\sigma = 1/t$ (so that $\mathrm{d}\sigma = -\mathrm{d}t/t^2$). Hence, on evaluating this easy integral, we find that

$$\mathrm{prob}(D|F, \sigma_\mathrm{o}, I) \;=\; \frac{1}{\sigma_\mathrm{o}\sqrt{2\pi}} \left[\frac{1 - e^{-R^2/2}}{R^2}\right], \tag{8.15}$$

where $R = (F-D)/\sigma_\mathrm{o}$ is the residual for the datum; it is plotted as a solid line in Fig. 8.2. Compared with the implicit Gaussian of a least-squares analysis, marked by a dashed line, where we believe that σ_o is the error-bar, $\mathrm{prob}(\sigma|\sigma_\mathrm{o}, I) = \delta(\sigma - \sigma_\mathrm{o})$, the pdf of eqn (8.15) is about 50% broader in the region of the central peak (where it's also lower by a factor of 2) and has slowly decaying Cauchy-like tails. It is the presence of the latter that is crucial in reducing the skewing effect of outliers, because the likelihood penalty for a large mismatch is then not so severe. For completeness, the dotted line in Fig. 8.2 shows the pdf that would have resulted if the marginal integral of eqn (8.14) had been evaluated with the Jeffreys' prior of eqn (8.12) with $\sigma_1 = 100\,\sigma_\mathrm{o}$.

If we treat all the data as above, and take the noise on the measurements as being independent, then, with the assignment of a uniform prior for the model parameters, the posterior pdf becomes

$$L = \log_\mathrm{e}\bigl[\mathrm{prob}(\boldsymbol{X}|\boldsymbol{D}, I)\bigr] = \mathrm{constant} + \sum_{k=1}^{N} \log_\mathrm{e}\left[\frac{1 - e^{-R_k^2/2}}{R_k^2}\right]. \tag{8.16}$$

While this is slightly more complicated than the least-squares case of eqns (8.1) and (8.5), it follows from all the same assumptions except that the quoted $\{\sigma_k\}$ are treated as lower bounds rather than being believed absolutely. The benefit of the extra effort involved in finding the X that maximizes the sum in eqn (8.16), over minimizing χ^2, is evident from the simple example of Fig. 8.1(b): the estimate of the gradient and intercept of the straight line is 12.0 ± 1.4 and 352.1 ± 7.0, respectively, instead of -5.3 ± 0.8 and 519.1 ± 3.8, and compares well with the true values of 10 and 350 used to generate the test data (corrupted intermittently by the addition of a large random positive number). The allowance for a global noise scaling in the least-squares framework, through the use of eqn (8.10), leads to estimates of -5.3 ± 10.4 and 519.1 ± 50.7; while this reduces the risk of over-interpreting the significance of the skewed result, it doesn't help in mitigating the effect of the outliers.

One of the drawbacks of the 'robust' formulation here, with respect to least-squares, is that the posterior pdf of eqn (8.16) is not guaranteed to be unimodal even when the model parameters are linearly related to the data. This is illustrated in Fig. 8.3, where the mean μ is estimated from a set of measurements (of the same quantity) containing one outlier. The solid line shows the posterior pdf of eqn (8.15), with $F_k = \mu$ and $\sigma_k = 1$, while the dotted line gives the equivalent result from the standard form of eqn (8.5).

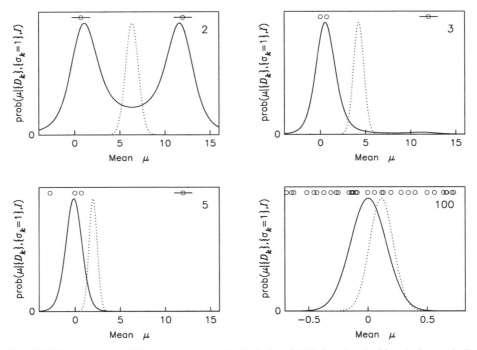

Fig. 8.3 The posterior pdf for the mean μ, after 2, 3, 5 and 100 data (marked by circles, and all with unit error-bar), when there is one outlier. The solid line is for the conservative formulation of eqn (8.16), whereas the dotted line is the result yielded by a standard least-squares analysis.

With just two data, our conservative formulation returns a bimodal distribution; this is entirely reasonable as, at this stage, we don't know which of the two inconsistent measurements is the outlier. By contrast, least-squares suggests that the true value is given by the arithmetic mean to a fair degree of reliability; this does not accord with common sense. When a third measurement is made, the pdf becomes almost unimodal. The small remnant peak is a warning that there might be two outliers! After all, we're treating all the data on an equal footing and haven't built into our analysis that there's only one corrupt point. As expected, the skewing effect of the rogue datum on least-squares is also seen to diminish with an increasing number of 'good' measurements.

8.3.2 The good-and-bad data model

Although eqn (8.16) provides valuable automatic protection against occasional quirky data, there is a price to pay: the uncertainties on the inferred model parameters are about 50% larger than they would be if we were able to believe the $\{\sigma_k\}$ as given. In the case of Fig. 8.1(a), for example, where there are no outliers, the conservative formulation returns estimates of 9.8 ± 1.2 and 351.4 ± 5.9 for the gradient and intercept of the straight line, respectively, whereas least-squares yields 9.8 ± 0.8 and 351.2 ± 3.8. Can we gain the benefit of a robust analysis without paying the related penalty?

Let's formulate the problem slightly differently. Rather than being very pessimistic, and saying that the noise associated with a measurement could only be worse than expected, we could allow just two possibilities: either everything behaved normally and the quoted error-bar is reasonable, or something went seriously wrong and the noise assessment should be scaled up by a large factor. In terms of eqn (8.13), this is equivalent to the assignment

$$\mathrm{prob}(\sigma|\sigma_\mathrm{o}, \beta, \gamma, I) = \beta\,\delta(\sigma - \gamma\sigma_\mathrm{o}) + (1-\beta)\,\delta(\sigma - \sigma_\mathrm{o}) , \qquad (8.17)$$

where $0 \leqslant \beta \ll 1$ and $\gamma \gg 1$. Substituting this into the marginal integral of eqn (8.15), along with a Gaussian pdf for $\mathrm{prob}(D|F, \sigma, I)$, we obtain

$$\mathrm{prob}(D|F, \sigma_\mathrm{o}, \beta, \gamma, I) = \frac{1}{\sigma_\mathrm{o}\sqrt{2\pi}} \left\{ \frac{\beta}{\gamma} \exp\left[-\frac{R^2}{2\gamma^2}\right] + (1-\beta)\exp\left[-\frac{R^2}{2}\right] \right\} , \qquad (8.18)$$

where $R = (F-D)/\sigma_\mathrm{o}$. With a uniform prior and the usual assumption of independent noise, the posterior pdf becomes

$$L = \mathrm{constant} + \sum_{k=1}^{N} \log_e\left[\frac{\beta}{\gamma}\,e^{-R_k^2/2\gamma^2} + (1-\beta)\,e^{-R_k^2/2} \right], \qquad (8.19)$$

where $L = \log_e\left[\mathrm{prob}(\boldsymbol{X}|\boldsymbol{D}, \beta, \gamma, I)\right]$, and reduces to standard least-squares in the limit $\beta \to 0$.

This formulation was first put forward by Box and Tiao (1968) and is conditional on a suitable choice for β and γ; namely, the frequency of the suspect measurements and the severity of their quirkiness. Although Box and Tiao preferred to examine the sensitivity of the results with respect to these parameters instead of marginalizing over them, estimates of 9.7 ± 0.8 and 351.3 ± 3.8 were obtained for the gradient and intercept of the straight line in Fig. 8.1(a) when β and γ were integrated out within the quadratic approximation of Section 3.2.1; thus, there appears to be no inferential penalty for encoding robustness in this way. The corresponding results from the data of Fig. 8.1(b) are also reasonable: 12.1 ± 0.9 and 351.0 ± 4.7. While this suggests that the formulation of eqn (8.19) is superior to that of eqn (8.16), we have found the latter to be more amenable to optimization with elementary search algorithms (e.g. simplex following an initial estimate from least-squares). The lack of extraneous parameters, such as β and γ, is both an advantage and a limitation of eqn (8.16).

8.3.3 The Cauchy formulation

Another useful alternative to the conservative formulation of eqn (8.13) is given by the assignment

$$\text{prob}(\sigma | \sigma_o, I) = \frac{2 \sigma_o}{\sqrt{\pi} \sigma^2} \exp\left(-\frac{\sigma_o^2}{\sigma^2}\right), \qquad (8.20)$$

which expresses an expectation that σ should be of the same order as σ_o, but could be either narrower or wider. Marginalizing σ according to eqn (8.14), using the substitution $\sigma = 1/t$, the likelihood contribution from datum D takes the Cauchy form

$$\text{prob}(D | F, \sigma_o, I) = \frac{1}{\sigma_o \pi \sqrt{2} \left[1 + (F - D)^2 / 2 \sigma_o^2\right]}.$$

This curve is shown as the dot–dashed line in Fig. 8.2. Once again, the tails decrease slowly to allow outliers, but the broadening of the central peak is reduced; in fact the curvature at the peak mimics that of the original Gaussian of width σ_o, albeit at a height reduced by a factor $\sqrt{\pi}$. With multiple independent data, and a uniform prior for the X, the posterior pdf takes the form

$$L = \log_e\left[\text{prob}(X | D, I)\right] = \text{constant} - \sum_{k=1}^{N} \log_e\left[1 + \frac{R_k^2}{2}\right]. \qquad (8.21)$$

This is equivalent to the standard least-squares eqn (8.5) when the residuals are small. For the data of Fig. 8.1(b), the gradient and intercept are estimated as 12.1 ± 1.2 and 350.4 ± 5.5.

More generally, it is permissible to assign any pdf to the likelihood, for example

$$\text{prob}(D | F, \sigma_o, \beta, I) \propto \left[1 + \frac{(F - D)^2}{2 \beta \sigma_o^2}\right]^{-\beta}.$$

In this formula, $\beta = 1$ is the Cauchy assignment, whilst $\beta \to \infty$ is the Gaussian. Instead of guessing a functional form for the noise, experience with other datasets from the same experimental apparatus might suggest, or even define, some other pdf specifically tuned to that apparatus. Anything goes, though the evidence values guide one towards a sensible choice.

8.4 Background removal

Although the data of Fig. 8.1(b) were used as a simple example of the outlier problem, they are reminiscent of the common case of a sharp signal sitting on a slowly varying background (albeit just a straight line). The correct way to deal with this situation is to evaluate the posterior pdf for the quantities of interest while marginalizing out the nuisance parameters, as was done in Section 4.2. This calculation can sometimes seem quite daunting, however, and we might seek a pragmatic separation between background and signal as a first step towards the analysis of the data. How can we use probability theory to help us do this?

Let's begin with the assumption of Gaussian noise, so that

$$\text{prob}(D \,|\, A, B, \sigma, I) \;=\; \frac{1}{\sigma\sqrt{2\pi}}\;\exp\!\left[-\frac{(A+B-D)^2}{2\,\sigma^2}\right], \qquad (8.22)$$

where σ is the expected mismatch between a single datum D and the corresponding sum of the background B and signal A. Since we cannot predict A without a model for the spectrum, it has to be marginalized out with a suitable prior $\text{prob}(A \,|\, I)$. If we knew only that A was positive and had a mean μ, for example, then the MaxEnt principle would lead to the exponential assignment of Section 5.3.1:

$$\text{prob}(A \,|\, \mu, I) \;=\; \frac{1}{\mu}\,\exp\!\left[-\frac{A}{\mu}\right] \qquad \text{for } A \geqslant 0, \qquad (8.23)$$

and zero otherwise. Substituting from eqns (8.22) and (8.23) into the A-integral of the joint pdf for A and D, $\text{prob}(D, A \,|\, B, \mu, \sigma, I) = \text{prob}(D \,|\, A, B, \sigma, I) \times \text{prob}(A \,|\, \mu, I)$, we obtain

$$\text{prob}(D \,|\, B, \mu, \sigma, I) \;=\; \frac{1}{\mu\,\sigma\sqrt{2\pi}} \int\limits_{0}^{\infty} e^{-A/\mu}\; e^{-(A+B-D)^2/2\sigma^2}\; \mathrm{d}A$$

$$= \; \frac{e^{\eta(R+\eta/2)}}{2\mu}\left[1 - \text{erf}\!\left(\frac{\eta+R}{\sqrt{2}}\right)\right], \qquad (8.24)$$

where the second line follows from the completion of the square for A in the exponent of the integrand and the definition of the *error function* in Appendix A.1, $\eta = \sigma/\mu$ and $R = (B-D)/\sigma$ is the background-based residual. This marginal likelihood is plotted as a dashed line in Fig. 8.4 for the case of $\eta = 0.1$; it is asymmetric, with a long tail allowing for large negative values of R.

Greater flexibility can be incorporated into the analysis by invoking the analogue of the good-and-bad data model, with $0 < \beta < 1$ being the chance that a datum contains a signal and $1-\beta$ that it's pure background. A weighted combination of the pdfs of eqn (8.22) with $A = 0$, giving the dotted Gaussian in Fig. 8.4, and eqn (8.24) yields

$$\text{prob}(D \,|\, B, \beta, \mu, \sigma, I) \;=\; \frac{\beta\, e^{\eta(R+\eta/2)}}{2\mu}\left[1 - \text{erf}\!\left(\frac{\eta+R}{\sqrt{2}}\right)\right] + \frac{(1-\beta)\, e^{-R^2/2}}{\sigma\sqrt{2\pi}}, \qquad (8.25)$$

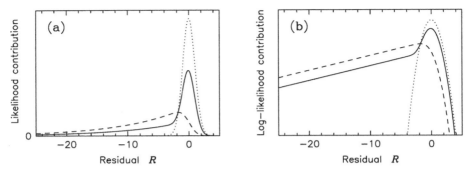

Fig. 8.4 (a) The likelihood contribution from a background-based residual $R = (B-D)/\sigma$. The dotted Gaussian is conditional upon there being purely background, or $A = 0$, while the dashed line is for the prior information $\langle A \rangle/\sigma = 10$; their weighted average, assuming a 50% chance of a signal, is shown by the solid curve. (b) The same pdfs plotted on a logarithmic axis.

and is marked by the solid line in Fig. 8.4 for $\beta = 0.5$ and $\eta = 0.1$. Assuming noise independence, the likelihood for a set of data D, given a slowly-varying model for the background described by a few parameters X, is a product of such terms for each of the measurements: $\mathrm{prob}(D\,|\,X, \beta, \mu, \sigma, I) = \prod \mathrm{prob}(D_k\,|\,B_k, \beta, \mu, \sigma_k, I)$ where R, σ and η in eqn (8.25) are replaced by $R_k = (B_k - D_k)/\sigma_k$ with $B_k = f(X, k)$ and $\eta_k = \sigma_k/\mu$. The nuisance parameters β and μ do need to be marginalized out, of course, to evaluate the posterior pdf for X. Carrying out the calculation for the data of Fig. 8.1 leads to virtually the same results as in Section 8.3.2; while there is no gain over eqn (8.21) in this case, we'd hope for an advantage in more trying circumstances. Examples of this type of analysis applied to real data can be found in Fischer *et al.* (2000) and in David and Sivia (2001).

8.5   Correlated noise: avoiding over-counting

One of the simplifying assumptions made throughout this book is that the uncertainty associated with one measurement is independent from that of any other. In the context of the Gaussian noise model, whose general multivariate form is given by

$$\mathrm{prob}(D\,|\,X, I) = \frac{1}{\sqrt{(2\pi)^N \det(\mathbf{C})}}\, \exp\!\left[-\tfrac{1}{2}\,(F-D)^{\mathrm{T}}\mathbf{C}^{-1}(F-D)\right], \qquad (8.26)$$

where $F - D$ is a column-vector of misfits ($F_k - D_k$ for $k = 1, 2, \dots N$), it is equivalent to taking the $(N \times N)$ covariance matrix $\mathbf{C}$ as being diagonal:

$$C_{kk'} = \begin{cases} \sigma_k^2 & \text{for } k = k', \\ 0 & \text{otherwise}, \end{cases} \qquad (8.27)$$

where $C_{kk'}$ is the kk'-element of $\mathbf{C}$. The inverse of this matrix is trivial, and leads to eqn (8.4) on substitution in eqn (8.26). When eqn (8.27) is not an adequate approximation to **C**, the data analysis prescription is still straightforward if we have a good estimate of its off-diagonal terms. All that changes is the definition of the misfit statistic:

$$\chi^2 = \sum_{k=1}^{N} \sum_{k'=1}^{N} (F_k - D_k) \left[\mathbf{C}^{-1}\right]_{kk'} (F_{k'} - D_{k'}) , \tag{8.28}$$

which reduces to eqn (8.1) for the case of eqn (8.27). If we suspect that correlations are not negligible but don't know their nature, how can we proceed?

8.5.1 Nearest-neighbour correlations

When faced with the situation where we lack important information, progress has to be made by invoking suitable assumptions; the simplest are best to begin with, but more sophistication can be added later if necessary. In this spirit, let's suppose that there is a characteristic nearest-neighbour correlation strength ϵ. This means that the expectation values of the pairwise products of misfits is constrained by

$$\left\langle (F_k - D_k)(F_{k'} - D_{k'}) \right\rangle = \begin{cases} \sigma_k^2 & \text{for } k = k', \\ \epsilon\, \sigma_k \sigma_{k'} & \text{for } |k - k'| = 1 , \end{cases} \tag{8.29}$$

where $-1 < \epsilon < 1$. Independence is equivalent to $\epsilon = 0$ and signifies that a higher than expected value of one measurement, for example, does not systematically correspond to a higher or lower value for the next. A tendency for same-signed deviations is indicated by $\epsilon > 0$, whereas oppositely-signed ones have $\epsilon < 0$.

The specification of eqn (8.29) does not define the covariance matrix completely, and so we are left with the task of assigning $\mathrm{prob}(\boldsymbol{D} \,|\, \boldsymbol{X}, I)$ subject to limited information. We saw how this could be done with the principle of MaxEnt in Chapter 5, when given testable constraints: the sole condition that $\left\langle (F_k - D_k)^2 \right\rangle = \sigma_k^2$ leads to eqn (8.4), while the additional knowledge of all the cross-correlations yields eqn (8.26). Maximizing the entropy of the likelihood function (relative to a uniform measure) subject to eqn (8.29), we obtain the Gaussian assignment of eqn (8.26) with a covariance matrix whose inverse is tridiagonal:

$$\left[\mathbf{C}^{-1}\right]_{kk'} = \begin{cases} \Lambda_k & \text{for } k = k', \\ \lambda_k & \text{for } |k - k'| = 1 , \end{cases} \tag{8.30}$$

where the $2N-1$ Lagrange multipliers, $\Lambda_1, \Lambda_2, \ldots, \Lambda_N, \lambda_1, \lambda_2, \ldots, \lambda_{N-1}$, have to be chosen to fulfil the requirements of eqn (8.29). The solution is found to be

$$\Lambda_k = \begin{cases} \dfrac{1}{(1 - \epsilon^2)\, \sigma_k^2} & \text{for } k = 1 \text{ or } N, \\[4mm] \dfrac{1 + \epsilon^2}{(1 - \epsilon^2)\, \sigma_k^2} & \text{for } 1 < k < N, \end{cases} \qquad \text{and} \quad \lambda_k = \dfrac{-\epsilon}{(1 - \epsilon^2)\, \sigma_k \sigma_{k+1}} , \tag{8.31}$$

with the resulting covariance matrix taking the simple form

$$C_{kk'} = \sigma_k \, \sigma_{k'} \, \epsilon^{|k-k'|} \, . \tag{8.32}$$

While this clearly satisfies eqn (8.29), its inverse relationship to eqns (8.30) and (8.31) is best verified by explicit multiplication.

Substituting for $\mathbf{C}^{-1}$ in eqn (8.26), with

$$\det(\mathbf{C}) = (\sigma_1 \sigma_2 \cdots \sigma_N)^2 \, (1 - \epsilon^2)^{N-1} \, , \tag{8.33}$$

and using the uniform prior of eqn (8.3), we find that the logarithm of the posterior pdf, $\log_e \big[\, \mathrm{prob}(\boldsymbol{X} \,|\, \boldsymbol{D}, \epsilon, I) \,\big]$, becomes

$$L = \mathrm{constant} - \frac{1}{2} \left[(N-1) \log_e(1 - \epsilon^2) + \frac{Q}{1 - \epsilon^2} \right] , \tag{8.34}$$

where Q consists of quantities quadratically related to the residuals $R_k = (F_k - D_k)/\sigma_k$. Explicitly,

$$Q = \chi^2 + \epsilon \left[\epsilon \, (\chi^2 - \phi) - 2\psi \right] , \tag{8.35}$$

where

$$\chi^2 = \sum_{k=1}^{N} R_k^2 \, , \quad \phi = R_1^2 + R_N^2 \quad \text{and} \quad \psi = \sum_{k=1}^{N-1} R_k \, R_{k+1} \, . \tag{8.36}$$

Reassuringly, eqn (8.34) reduces to the least-squares form of eqn (8.5) when $\epsilon = 0$. It is interesting to note how MaxEnt has filled out the whole of the covariance matrix, with a power-law decay $\epsilon^{|k-k'|}$, even though the constraints were only on the variance and first off-diagonal terms. Not only is the result of eqn (8.32) simple, but one that we might have tried intuitively; indeed, Bernardo and Smith (1994) give it as a possibility without offering any theoretical justification.

When faced with the situation where the quoted error-bars 'don't look right', our first inclination would be to suspect that they have been wrongly scaled: the σ_k should really have been $\gamma \sigma_k$, for all k, where the value of the constant γ is not known. If this uncertainty is taken into account as well as allowing for the possibility of correlations, then an analysis similar to that in Section 8.2 leads to a marginal posterior pdf (with γ integrated out) analogous to eqn (8.10):

$$L = \mathrm{constant} - \frac{1}{2} \left[N \log_e(Q) - \log_e(1 - \epsilon^2) \right] , \tag{8.37}$$

where $L = \log_e \big[\, \mathrm{prob}(\boldsymbol{X} \,|\, \boldsymbol{D}, \epsilon, I) \,\big]$, and Q is given by eqns (8.35) and (8.36).

8.5.2 An elementary example

The easiest case to analyse is the situation considered in Section 2.3: Given a set of N measurements of a quantity μ, $\{x_k\}$ all with an error-bar σ, what is our best estimate μ_o and how confident are we of this prediction? The answer derived was the arithmetic mean of the data, with an uncertainty of $\sigma/\sqrt{N}$. How would eqns (2.28) and (2.29) change if the measurements were subject to nearest-neighbour correlations of strength ϵ? Well, substituting $F_k = \mu$, $D_k = x_k$ and $\sigma_k = \sigma$ in eqns (8.34)–(8.36), the optimality condition $\mathrm{d}L/\mathrm{d}\mu\big|_{\mu_o} = 0$ yields the solution

$$\mu_o = \frac{1}{N - \epsilon(N-2)} \left[x_1 + x_N + (1-\epsilon) \sum_{k=2}^{N-1} x_k \right].$$ (8.38)

Its reliability, calculated by taking the square root of the inverse of $d^2 L / d\mu^2$, is given by

$$\mu = \mu_o \pm \sigma \sqrt{\frac{1 + \epsilon}{N(1-\epsilon) + 2\epsilon}} \approx \mu_o \pm \frac{\sigma}{\sqrt{N}} \sqrt{\frac{1 + \epsilon}{1 - \epsilon}},$$ (8.39)

where the simplification on the right holds for moderately large N and ϵ not too close to unity, when eqn (2.28) is also a good approximation to μ_o.

As required, eqns (8.38) and (8.39) reduce to eqns (2.28) and (2.29) when $\epsilon = 0$. In the limit of positive correlation, when $\epsilon \to 1$, all the data become identical. With $x_k = x$, for all k, we obtain $\mu = x \pm \sigma$; this is independent of N, as expected. In the limit of anti-correlation, when $\epsilon \to -1$, the data alternate between two values spaced equally around the true μ; eqns (8.38) and (8.39) then return the correct μ without uncertainty! Having passed these basic tests, we can feel more confident that our analysis will yield sensible results for intermediate levels of correlation. Even the strange asymmetry in the weighting of the measurements in eqn (8.38) can be understood by recognizing that the first and last points have only one neighbour, whereas all the others have two.

Although we tend to think of correlations in the noise as a bad thing, because they reduce the effective number of 'independent' measurements, quantified approximately by the $(1+\epsilon)/(1-\epsilon)$ factor in eqn (8.39), the striking difference in the behaviour of the reliability of μ_o in the two limits of $\epsilon \to \pm 1$ warns us that this view may be overly simplistic. If we were fitting a straight line to a pertinent set of data where the noise was subject to serious positive correlation, say, then this would have a detrimental effect on our estimate of the intercept but a beneficial one on the gradient. This is because, in the limit of $\epsilon \to 1$, all the measurements would deviate from the straight line by the same fraction of their error-bars; with $\sigma_k = \sigma$, the data would exactly match the true gradient but be displaced vertically by a random amount (determined by σ).

8.5.3 Time series

A treatment of correlated noise inherently assumes an ordering of the data; otherwise, terms such as 'nearest-neighbour' have little meaning. Since measurements which are a function of (increasing) time t are naturally sequential, $D_k = D(t_k)$ with $t_k > t_{k-1}$, data where the ordering is important are often referred to as a *time series*. Dependence of a measurement on the preceding ones is called a *Markov* process, and the length of the backwards linkage is the 'order of the chain'. Independence is equivalent to zeroth order,

$$\text{prob}(D_k | D_{k-1}, D_{k-2}, \ldots, D_1, X, I) = \text{prob}(D_k | X, I),$$ (8.40)

where X is the vector of parameters for the model describing the systematic variation of the data, and the (unqualified) term 'Markov chain' is most frequently used to denote a first order *Brownian* motion-like process:

$$\text{prob}(D_k | D_{k-1}, D_{k-2}, \ldots, D_1, X, I) = \text{prob}(D_k | D_{k-1}, X, I).$$ (8.41)

An example of a time series is shown in Fig. 8.5(a). It relates to a *molecular dynamics* simulation of 256 water molecules at room temperature; the details can be found in Refson (2000) and an introduction in Allen and Tildesley (1987). Many physical properties of the system, such as the potential energy, the rotational and translational kinetic energies, the pressure, the dipole moment, the stress tensor and so on, can be computed

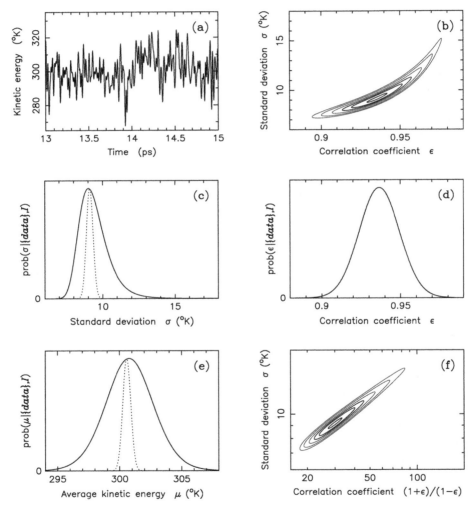

Fig. 8.5 (a) The evolution in the kinetic energy of a model water system, sampled at intervals of 0.0025 ps. (b) The joint posterior pdf for the correlation coefficient, ϵ, and the standard deviation, σ, of the fluctuations: $\text{prob}(\epsilon, \sigma \,|\, \{data\}, I)$. (c)–(e) The marginal posterior pdfs for ϵ, σ and the average kinetic energy μ; the dotted lines show the equivalent results assuming independence ($\epsilon = 0$). (f) The same as (b), except that axes are logarithmic in σ and the ratio $(1+\epsilon)/(1-\epsilon)$.

as a function of time. Figure 8.5(a) shows 800 sequential samples of the instantaneous kinetic energy, plotted in terms of equivalent degrees Kelvin, over a 2 picosecond interval. If the data are regarded as fluctuations about a uniform value μ, as in Section 8.5.2, then the joint posterior pdf for the correlation strength ϵ, the error-bar σ and μ is given by

$$L = L_p - \frac{1}{2}\left[(N-1)\log_e(1-\epsilon^2) + 2N\log_e(\sigma) + \frac{Q}{\sigma^2(1-\epsilon^2)}\right], \qquad (8.42)$$

where $R_k = \mu - D_k$ in eqn (8.36), L_p is a constant plus the logarithm of the prior pdf for μ, σ and ϵ, and $L = \log_e\left[\text{prob}(\mu, \sigma, \epsilon \,|\, D, I)\right]$; this is just eqn (8.34) with all the factors of σ retained explicitly. To obtain the posterior pdf for σ and ϵ, the exponential of L has to be integrated with respect to μ. This marginalization can be done analytically because, for a given σ and ϵ (and a uniform μ-prior), the task of optimizing μ is linear. Hence,

$$\text{prob}(\sigma, \epsilon \,|\, D, I) \propto \text{prob}(\sigma, \epsilon \,|\, I) \times \exp(L_0)\left[-\frac{d^2L}{d\mu^2}\right]^{-1/2}, \qquad (8.43)$$

where L_0 is the value of eqn (8.42) when $dL/d\mu = 0$. Carrying out this calculation with a uniform $\text{prob}(\sigma, \epsilon \,|\, I)$ yields the pdf in Fig. 8.5(b); it has been contoured at 0.5, 2.5, 12.5, 30, 60 and 90% of the maximum. The integrations of Fig. 8.5(b) along the two coordinate axes give the marginal posterior pdfs for σ and ϵ shown in Figs 8.5(c) and (d). The latter confirms the presence of significant nearest-neighbour correlation, with $\epsilon \approx 0.94$, whereas the dotted line in the former warns us about the danger of over confidence in assuming independence. This tendency can also be seen in Fig. 8.5(e), where the error-bar for μ is about five times too small if we take ϵ as being zero rather than marginalizing it out; this is consistent with the $(1+\epsilon)/(1-\epsilon)$ term in eqn (8.39). As always, however, we must not forget that our conclusions are conditional on the assumptions that underlie the analysis.

The assignment of $\text{prob}(\sigma, \epsilon \,|\, I) \propto [\sigma(1-\epsilon^2)]^{-1}$ might be a better choice than a naïve uniform prior. While the scale parameter argument for $1/\sigma$ is given in Section 5.1.2, $(1-\epsilon^2)^{-1}$ is a suggestion based on the observation that the correlation ellipse of Fig. 3.6 changes rapidly as $\epsilon \to \pm 1$. The potential benefit of our proposal, which is equivalent to having a uniform prior in $\log(\sigma)$ and $\log[(1+\epsilon)/(1-\epsilon)]$, can be seen from the more convenient shape of the posterior pdf in Fig. 8.5(f). When this prior was used for the data of Fig. 8.5(a), the results were essentially unchanged.

8.6 Log-normal: least-squares for magnitude data

When discussing location and scale parameters in Section 5.1.2, we saw that it was more natural to work in logarithmic coordinates for the latter. This simply reflects the fact that it's the relative change that is most important for such quantities. For example, economic changes are usually given in percentage terms. Likewise, it may be preferable to work with fractional reliabilities for magnitude-type data: a length is 50 m to within 2%, say, instead of ± 1 m. Although this might sound strange, and is unnecessary for

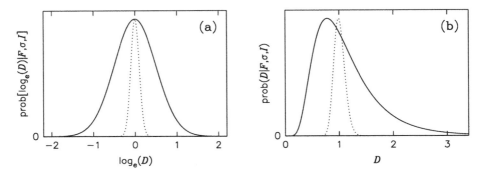

Fig. 8.6 (a) The Gaussian likelihood for the logarithm of datum D given an expected value of $F=1$ and a relative error σ of 10% (dotted) and 50%. (b) The equivalent log-normal pdf for D.

many situations, it has a bearing on the precise choice of our likelihood function. If we let $y = \log_e(x)$, so that

$$\left\langle (\delta y)^2 \right\rangle = \left\langle \left(\frac{\delta x}{x} \right)^2 \right\rangle ,$$

then a constraint on the relative error in x is equivalent to an absolute one in y. The MaxEnt principle then leads us to assign a Gaussian pdf for $\log_e(x)$ or, through a change of variables with eqn (3.77), a _log-normal_ distribution in x:

$$\text{prob}(x|x_o, \sigma) = \frac{1}{x \, \sigma \sqrt{2\pi}} \, \exp\left\{ -\frac{[\log_e(x/x_o)]^2}{2\sigma^2} \right\} , \qquad (8.44)$$

for positive x, and zero otherwise, where $x_o > 0$ is the _median_ value of x with relative variance $\sigma^2 = \left\langle (\delta x/x)^2 \right\rangle$; two cases are illustrated in Fig. 8.6. Since the assumption of Gaussian noise is central to the use of least-squares, and a poor approximation to eqn (8.44) when σ is more than about 10%, it's better to work with the logarithm of magnitude data.

9 Nested sampling

The resurgence of interest in the Bayesian approach to data analysis has been driven, in practical terms, partly by the rapid development of computer hardware and partly by the advent of larger-scale problems. This chapter discusses some of the modern numerical techniques that are useful for doing Bayesian calculations when analytical approximations are inadequate; in particular, it focuses on the novel idea of *nested sampling*.

9.1 Introduction: the computational problem

In this chapter we are concerned, not so much with the probabilistic formalism, but with how to compute it once the hypothesis space and the data have been assigned. To orient ourselves, let's expand the joint probability for the parameters x of a given model and the data D in the two different ways allowed by the product rule:

$$
\begin{array}{ccccc}
\mathrm{prob}(\boldsymbol{D}|\boldsymbol{x}, I)\ \mathrm{prob}(\boldsymbol{x}|I) & = & \mathrm{prob}(\boldsymbol{x}, \boldsymbol{D}|I) & = & \mathrm{prob}(\boldsymbol{D}|I)\ \mathrm{prob}(\boldsymbol{x}|\boldsymbol{D}, I) \\
\mathcal{L}(\boldsymbol{x}) \qquad \pi(\boldsymbol{x}) & = & \cdots & = & Z \qquad\quad P(\boldsymbol{x}) \\
\text{Likelihood} \times \text{Prior} & = & \text{Joint} & = & \text{Evidence} \times \text{Posterior} \\
\text{inputs} & \Longrightarrow & \cdots & \Longrightarrow & \text{outputs}
\end{array}
$$

from which Bayes' theorem follows trivially. The inputs to our computation are the prior $\pi(\boldsymbol{x})$ and the likelihood $\mathcal{L}(\boldsymbol{x})$, whilst the desired outputs are the evidence Z and the posterior $P(\boldsymbol{x})$; the compressed notation emphasizes that our focus is on algorithmic methods rather than probabilistic foundation.

The normalization requirement for probability distributions means that the cumulant masses of both the prior and the posterior are unity,

$$
\iint \cdots \int \pi(\boldsymbol{x})\, \mathrm{d}\boldsymbol{x} \ = \ 1 \quad \text{and} \quad \iint \cdots \int P(\boldsymbol{x})\, \mathrm{d}\boldsymbol{x} \ = \ 1 . \tag{9.1}
$$

In the terminology of the subject, *mass* denotes an accumulated amount of probability, and the pdf (probability density function) is its differential $\mathrm{d}(mass)/\mathrm{d}(volume\ \text{of}\ \boldsymbol{x})$. It is the second condition in eqn (9.1) that lets us separate the amount Z of joint distribution from the shape P, through

$$
Z \ = \ \iint \cdots \int \mathcal{L}(\boldsymbol{x})\, \pi(\boldsymbol{x})\, \mathrm{d}\boldsymbol{x} , \tag{9.2}
$$

whence

$$
P(\boldsymbol{x}) \ = \ \frac{\mathcal{L}(\boldsymbol{x})\, \pi(\boldsymbol{x})}{Z} . \tag{9.3}
$$

Our aim is to evaluate these outputs for problems which are too large for brute-force enumeration of 'all' $\boldsymbol{x}$, or for adequate approximation by any convenient algebraic form.

In this context, it is worth remembering that a space of high dimensionality has an exponentially large number of constituent regions: at a resolution of 1 part in R, a space of dimension N has R^N elements. Hence even quite moderate N can defeat the brute-force approach, and have too much freedom for useful analytical approximation.

9.1.1 Evidence and posterior

What we perhaps cannot do in practice, we can nevertheless contemplate doing in principle. Just as we can think of evaluating Z in eqn (9.2) by the direct summation of small elements, so we can imagine sorting these elements into decreasing order of likelihood value. Taking the example of a two-dimensional model, where $\boldsymbol{x} = (x_1, x_2)$, the 4×4 table of likelihood values

$$
\mathcal{L} =
\begin{array}{|c|c|c|c|}
\hline
0 & 8 & 15 & 3 \\
\hline
11 & 24 & 22 & 10 \\
\hline
19 & 30 & 26 & 16 \\
\hline
9 & 23 & 18 & 6 \\
\hline
\end{array}
$$

would sort into a 16-cell vector:

$$\mathcal{L} = (30, 26, 24, 23, 22, 19, 18, 16, 15, 11, 10, 9, 8, 6, 3, 0).$$

The accumulation of such elements in the context of a general $\boldsymbol{x}$ lets us acquire the function

$$\xi(\lambda) = \text{proportion of prior with likelihood greater than } \lambda. \tag{9.4}$$

More formally,

$$\xi(\lambda) = \underset{\mathcal{L}(\boldsymbol{x}) > \lambda}{\iint \cdots \int} \pi(\boldsymbol{x}) \, \mathrm{d}\boldsymbol{x} \,, \tag{9.5}$$

in which the element of prior mass is $\mathrm{d}\xi = \pi(\boldsymbol{x}) \, \mathrm{d}\boldsymbol{x}$. Dimensionality is not a problem: if each of the N components of $\boldsymbol{x}$ is stored to an accuracy of 1 part in R, then the single

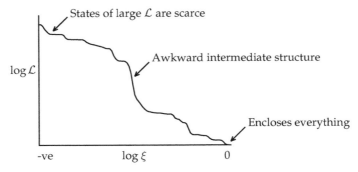

Fig. 9.1 A sorted likelihood function, on logarithmic axes: ξ is the proportion of the prior mass in which the model parameters yield a likelihood value greater than a threshold $\mathcal{L}$.

coordinate ξ should be stored to 1 part in R^N, requiring the same number $N \log_2 R$ of bits as for $\boldsymbol{x}$, implying no loss of information. At any degree of decomposition, there is a 1:1 correspondence between $\boldsymbol{x}$ and ξ. At least, that's the theory. In practice, we can store ξ to ordinary arithmetical precision, and in continuous problems leave ambiguous ties of likelihood to look after themselves. In discrete problems, where a substantial fraction of the prior mass can all be assigned exactly the same likelihood value, ties can be resolved by adding a small amount of random jitter to $\mathcal{L}$. This imposes a ranking even when none existed before.

Because the restriction on likelihood becomes tighter as λ increases, ξ is a decreasing (with jitter, strictly decreasing) function of likelihood limit λ, but it need not have any other nice property. The extreme values are $\lambda \geqslant 0$ at $\xi = 1$, because likelihood values cannot be negative, and $\lambda = \mathcal{L}_{\max}$ (if the maximum exists) at $\xi = 0$. The differential $-\mathrm{d}\xi/\mathrm{d}\lambda$, insofar as the small-scale limit exists, is the density of states; specifically, the density of prior mass with respect to likelihood.

It's actually more convenient to use the equivalent inverse form of eqn (9.4) or (9.5), in which the enclosed prior mass ξ is the primary variable and $\mathcal{L}$ the subsidiary, $\mathcal{L}(\xi)$ being defined by

$$\mathcal{L}(\xi(\lambda)) \equiv \lambda. \tag{9.6}$$

Note that $\mathcal{L}(\xi)$, having scalar argument ξ, should not be confused with $\mathcal{L}(\boldsymbol{x})$, having vector argument $\boldsymbol{x}$. In accordance with modern computing practice, we *overload* the symbol $\mathcal{L}$ with formally different functions according to the argument type, and also use $\mathcal{L}$ for the common value that these functions take. A typically awkward likelihood is sketched in Fig. 9.1, using logarithmic axes because even after taking logs the scales of these axes could easily be thousands or millions in a respectably large application.

The likelihood function $\mathcal{L}(\xi)$ underlies both the evidence Z and the posterior P. To see this, remember that each small element $\mathrm{d}\xi$ came from a source volume of that same

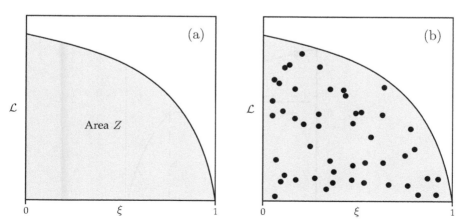

Fig. 9.2 (a) Likelihood function with area Z, not to scale since most of the area is likely to be found at invisibly tiny values of ξ. (b) Posterior samples are randomly scattered over the area Z.

prior mass $\pi(\boldsymbol{x})\,\mathrm{d}\boldsymbol{x}$. Hence the evidence of eqn (9.2), being a sum over these elements, is simply the enclosed area as shown in Fig. 9.2(a):

$$Z = \int\limits_{0}^{1} \mathcal{L}(\xi)\,\mathrm{d}\xi\,. \tag{9.7}$$

Any point taken randomly from this area, as illustrated in Fig. 9.2(b), yields a random sample $\tilde{\xi}$ from the posterior distribution

$$P(\xi) = \frac{\mathcal{L}(\xi)}{Z} \tag{9.8}$$

which, by the same argument, gives equivalently a random sample $\tilde{x}$ from the posterior $P(\boldsymbol{x})$. Hence the sorted likelihood function $\mathcal{L}(\xi)$ is the key to obtaining both the evidence Z and the posterior P (as a set of random samples).

9.2   Nested sampling: the basic idea

The new technique of *nested sampling* (Skilling 2004) tabulates the sorted likelihood function $\mathcal{L}(\boldsymbol{x})$ in a way that itself uses Monte Carlo methods. The technique uses a collection of n *objects* x, randomly sampled with respect to the prior π, but also subject to an evolving constraint $\mathcal{L}(\boldsymbol{x}) > \mathcal{L}^*$ preventing the likelihood from exceeding the current limiting value $\mathcal{L}^*$. We assume that we are able to generate such objects, and will discuss the accomplishment of this in Section 9.3.

In terms of ξ, the objects are uniformly sampled subject to the constraint $\xi < \xi^*$, where ξ^* corresponds to $\mathcal{L}^*$; this is illustrated in Fig. 9.3. At the outset, sampling is uniform over the entire prior, meaning that $\xi^* = 1$ and $\mathcal{L}^* = 0$. The idea is then to iterate inwards in ξ and correspondingly upwards in $\mathcal{L}$, in order to locate and quantify the tiny region of high likelihood where most of the joint distribution is to be found.

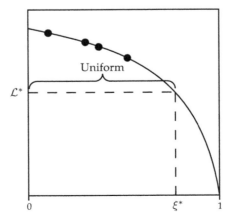

Fig. 9.3 Four objects ($n=4$) sampled uniformly in $\xi < \xi^*$, or equivalently in $\mathcal{L} > \mathcal{L}^*$.

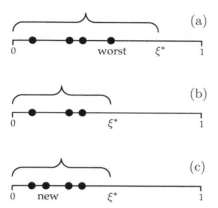

Fig. 9.4 An iteration replaces the worst object with a new one inside the shrunken domain.

9.2.1 Iterating a sequence of objects

On entry, an iteration holds n objects restricted to $\xi < \xi^*$, as shown in Fig. 9.4(a). The worst of these, being the one with smallest likelihood and hence largest ξ, is selected. Located at the largest of n numbers uniformly distributed in $(0, \xi^*)$, it will lie about one part in n less than ξ^*. More technically, the shrinkage ratio $t = \xi/\xi^*$ is distributed as

$$\mathrm{prob}(t) = n\,t^{n-1}, \qquad (9.9)$$

with mean and standard deviation

$$\log t = (-1 \pm 1)/n . \qquad (9.10)$$

Iteration proceeds by using the worst object's $(\xi, \mathcal{L})$ as the new $(\xi^*, \mathcal{L}^*)$. Meanwhile, the worst object, no longer obeying the constraint, is discarded. There are now $n-1$ surviving objects, still distributed uniformly over ξ but confined to a shrunken domain bounded by the new constraint ξ^*; this is illustrated in Fig. 9.4(b). The new domain is nested within the old, hence the name 'nested sampling'. The next step is to generate a replacement object, sampled uniformly over the prior but constrained within this reduced domain. For now, we assume that we are able to do this. Having done it, the iteration again holds n objects restricted to $\xi < \xi^*$, as in Fig. 9.4(c), just like on entry except for the 1-part-in-n shrinkage. The loop is complete, and the next iteration can be started.

Successive iterations generate a sequence of discarded objects on the edges of progressively smaller nested domains. At iterate k,

$$\mathcal{L}_k = \mathcal{L}^* \quad \text{and} \quad \xi_k = \xi^* = \prod_{j=1}^{k} t_j, \qquad (9.11)$$

in which each shrinkage ratio t_j is independently distributed with the pdf of eqn (9.9) with the statistics of eqn (9.10). It follows that

$$\log \xi_k = (-k \pm \sqrt{k})/n . \qquad (9.12)$$

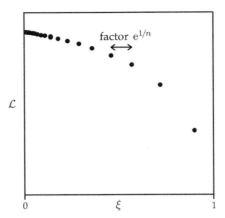

Fig. 9.5 The sorted likelihood function, $\mathcal{L}(\xi)$, is generated as a discrete sequence.

Ignoring uncertainty for a moment, we can proclaim each $\log t$ to be $-1/n$ so that $\xi_k = \exp(-k/n)$, and the sequence then tabulates $\mathcal{L}(\xi)$ just as we require. Everything we want is now available, as shown in Fig. 9.5.

The evidence of eqn (9.7) is evaluated by associating with each object in the sequence a width $h = \Delta\xi$, and hence a vertical strip of area $A = h\mathcal{L}$, whence

$$Z \approx \sum_k A_k, \quad \text{where } A_k = h_k \mathcal{L}_k. \tag{9.13}$$

Here, the simplest assignment of the width is

$$h_k = \xi_{k-1} - \xi_k. \tag{9.14}$$

One can try to be more accurate by using the trapezoid or such rule instead, but the uncertainties in ξ tend to overwhelm these minor variations of implementation. Along with the evidence, the *information*, or negative entropy,

$$\mathcal{H} = \int P(\xi) \log[P(\xi)] \, d\xi \approx \sum_k \frac{A_k}{Z} \log\left[\frac{\mathcal{L}_k}{Z}\right]$$

is available as the natural logarithmic measure of the prior-to-posterior shrinkage (as discussed in Section 7.4, but here using natural logarithms). Very roughly,

$$\mathcal{H} \approx (\# \text{ active components in data}) \times \log(\text{signal/noise ratio}). \tag{9.15}$$

9.2.2 Terminating the iterations

The usual behaviour of the areas A is that they start by rising, with the likelihood $\mathcal{L}$ increasing faster than the widths h decrease. The more important regions are being found. At some point, $\mathcal{L}$ flattens off sufficiently that decreasing width dominates increasing likelihood, so that the areas pass across a maximum and start to fall away. Most of the total area is usually found in the region of this maximum, which occurs in the region of $\xi \approx e^{-\mathcal{H}}$. Very roughly, the width of this region is

$$\Delta\left(\log\xi\right) \approx \sqrt{\text{\# active components in data}} , \qquad (9.16)$$

in accordance with the '$\chi^2 = N \pm \sqrt{2N}$' folklore surrounding Gaussian data from eqn (3.83). Remembering that $\xi_k \approx e^{-k/n}$ suggests a termination condition

'continue iterating until the count k significantly exceeds $n\mathcal{H}$'

which conveniently expresses the general aim that a nested-sampling calculation should be continued until most of Z has been found. Of course, $\mathcal{H}$ is here the current evaluate from the previous k iterates.

Unfortunately, we can offer no rigorous criterion to ensure the validity of any such termination condition. It is perfectly possible for the accumulation of Z to flatten off, apparently approaching a final value, whilst yet further inward there lurks a small domain in which the likelihood is sufficiently large to dominate the eventual results. Exactly this happens in Section 9.6.2, where termination is crudely imposed by stopping after a sufficient number of iterates. Termination remains a matter of user judgment about the problem in hand, albeit with the aim of effectively completing the accumulation of Z.

9.2.3 Numerical uncertainty of computed results

The $n\mathcal{H}$ iterates taken to reach the dominating areas are random, and according to eqn (9.12) are subject to a standard deviation uncertainty $\sqrt{n\mathcal{H}}$. Correspondingly, the accumulated values of $\log\xi$ are subject to a standard deviation uncertainty $\sqrt{n\mathcal{H}}/n$. There will be some internal variation as well, as the dominating region is traversed, but the major uncertainty will be a scale factor caused by the shift in logarithm. This shift is transmitted to the evidence Z through eqn (9.13), so that $\log Z$ too has standard deviation uncertainty $\sqrt{n\mathcal{H}}/n$. Putting the results together,

$$\log Z \approx \log\left(\sum_k A_k\right) \pm \sqrt{\frac{\mathcal{H}}{n}} . \qquad (9.17)$$

Incidentally, this uncertainty range is best expressed, as shown, in terms of $\log Z$; translating it back to a mean and standard deviation of Z itself just looks misleading whenever the range of the logarithm is substantial. The choice matters. For example, if the evidence values for hypotheses C and D were quoted as $\log_e Z_C = -1000$ and $\log_e Z_D = -1010$, one would be inclined to favour C by a factor of e^{10}. Yet if the estimates were accompanied by uncertainties of ±100, the comparison would effectively be just a toss-up. The difference would then be $\log_e Z_C - \log_e Z_D = 10 \pm 100\sqrt{2}$, for which the chance of C bettering D is a mere 53%. If the difference was important, more computation would be needed to reduce the uncertainty.

Also, remember that Z is dimensional, having dimensions inverse to the data. For example, if the relevant data were 157 measurements of length and 23 of time, the dimension of Z would be $m^{-157}s^{-23}$, so that Z should be quoted in a form akin to

$$\log_e\left(Z \, / \, m^{-157}s^{-23}\right) = \text{estimate} \pm \text{numerical uncertainty} .$$

A more sophisticated approach to the uncertainty uses the known distribution eqn (9.9) of each shrinkage ratio t. If we knew the t's, we would know all the ξ's, thence all the widths, and (apart from minor systematic errors in the numerical integration) the evidence and everything else. In other words, we know $Z(t)$ as a function of the t's:

$$t \overset{(9.11)}{\Longrightarrow} \xi \overset{(9.14)}{\Longrightarrow} h \overset{(9.13)}{\Longrightarrow} A \overset{(9.13)}{\Longrightarrow} Z.$$

So all we need to do is sample the t's a dozen times or more in order to obtain a corresponding list of evidence values. From such a list, we extract statistics (mean and standard deviation, quantiles or whatever) just as from any other set of Monte Carlo samples. The illustrative approximation of eqn (9.17) should be good enough for many purposes, but the sophisticated option is more defensible professionally.

Historically, it has not been usual to present an uncertainty range along with $\log Z$. There has, perhaps, been a feeling that numerical calculations should be accurate. However, there are inevitable uncertainties associated with nested sampling, as with any other numerical method, and it is honest to acknowledge them. The uncertainty diminishes as the square root of the amount of computation (here n) that was allotted, just as for other Monte Carlo methods.

9.2.4   Programming nested sampling in 'C'

Although small, the following 'C' main program incorporates the above ideas and should suffice for many applications. It is protected against over/underflow of exponential quantities such as the likelihood $\mathcal{L}^*$ by storing them as logarithms (as in the variable `logLstar`), adding those values through the PLUS macro, and multiplying them through summation. Rather than attempting greater sophistication, the program uses the simple proclamation of steady compression by $\log(t) = -1/n$ each step. The corresponding step-widths h are also stored as logarithms, in `logwidth`.

```
//                      NESTED SAMPLING MAIN PROGRAM
//            (GNU GENERAL PUBLIC LICENSE  software: © Sivia and Skilling 2006)
#include <stdio.h>
#include <stdlib.h>
#include <math.h>
#include <float.h>
#define UNIFORM ((rand()+0.5)/(RAND_MAX+1.0))   // Uniform inside (0,1)
#define PLUS(x,y) (x>y ? x+log(1+exp(y-x)) : y+log(1+exp(x-y)))
                                // logarithmic addition log(exp(x)+exp(y))
/*_____*/
#include "apply.c"            // Application code, setting int n, int MAX,
                              // struct Object, void Prior,
                              // void Explore, void Results.
/*_____*/
int main(void)
{
    Object Obj[n];           // Collection of n objects
    Object Samples[MAX];     // Objects stored for posterior results
    double logwidth;         // ln(width in prior mass)
    double logLstar;         // ln(Likelihood constraint)
    double H     = 0.0;      // Information, initially 0
    double logZ =-DBL_MAX;// ln(Evidence Z, initially 0)
```

```
    double logZnew;          // Updated logZ
    int    i;                // Object counter
    int    copy;             // Duplicated object
    int    worst;            // Worst object
    int    nest;             // Nested sampling iteration count
// Set prior objects
    for( i = 0; i < n; i++ )
        Prior(&Obj[i]);
// Outermost interval of prior mass
    logwidth = log(1.0 - exp(-1.0 / n));
// NESTED SAMPLING LOOP _____
    for( nest = 0; nest < MAX; nest++ )
    {
// Worst object in collection, with Weight = width * Likelihood
        worst = 0;
        for( i = 1; i < n; i++ )
            if( Obj[i].logL < Obj[worst].logL )  worst = i;
        Obj[worst].logWt = logwidth + Obj[worst].logL;
// Update Evidence Z and Information H
        logZnew = PLUS(logZ, Obj[worst].logWt);
        H = exp(Obj[worst].logWt - logZnew) * Obj[worst].logL
          + exp(logZ - logZnew) * (H + logZ) - logZnew;
        logZ = logZnew;
// Posterior Samples (optional)
        Samples[nest] = Obj[worst];
// Kill worst object in favour of copy of different survivor
        do copy = (int)(n * UNIFORM) % n;  // force 0 <= copy < n
        while( copy == worst && n > 1 );   // don't kill if n is only 1
        logLstar = Obj[worst].logL;        // new likelihood constraint
        Obj[worst] = Obj[copy];            // overwrite worst object
// Evolve copied object within constraint
        Explore(&Obj[worst], logLstar);
// Shrink interval
        logwidth -= 1.0 / n;
    } // _____ NESTED SAMPLING LOOP (might be ok to terminate early)
// Exit with evidence Z, information H, and optional posterior Samples
    printf("# iterates = %d\n", nest);
    printf("Evidence: ln(Z) = %g +- %g\n", logZ, sqrt(H/n));
    printf("Information: H = %g nats = %g bits\n", H, H/log(2.));
    Results(Samples, nest, logZ);          // optional
    return 0;
}
```

Accompanying the main program, an application module is to be '#include'd as file 'apply.c'. This needs to #define the number n of objects, and the number MAX of iterates. The application module also needs to define the Object structure needed to hold a sample, and procedures `Prior` for setting up a prior sample, `Explore` for finding a new sample within the current likelihood constraint and optionally `Results` for calculating such posterior results as might be wanted. It is open to the main program to terminate early if some suitable condition is satisfied, though that is not implemented here. With the application module programmed, nested sampling is complete. From the user, it requires judgment about the number of objects needed and the termination decision.

9.3 Generating a new object by random sampling

To complete the account of nested sampling, we must now discuss how to accomplish the required task of generating a new object, and this too involves randomness.

Computers are causal devices that perform definitive reproducible calculations, so are not random at all. What is described as a *random number* is actually a number generated by an algorithm that is carefully constructed to give successive outputs that are un-predictable in the absence of knowledge of the generator itself — that's what 'random' means in Bayesian computation. Such algorithms (Press *et al.* 1986) provide random integers in some range, most conveniently 0 to $2^{32} - 1$, or 0 to $2^{64} - 1$, whose values are unpredictable by the ignorant. It is easy to rescale such an integer into the standard range used in floating-point work:

$$u = \texttt{Uniform}(0,1) \iff \text{prob}(u \,|\, \text{ignorance of generator}) = \begin{cases} 1 & \text{for } 0 < u < 1, \\ 0 & \text{otherwise}. \end{cases}$$

As a professional detail, a good generator should produce odd multiples of 2^{-24} for ordinary IEEE 23-bit-mantissa floating-point format, or odd multiples of 2^{-53} for a 52-bit-mantissa format. Such numbers are *a-priori*-equivalent and symmetrically arranged in the strict interior of $(0,1)$, with no stray anomalies such as an exact 0 or 1.

Initializing a nested-sampling calculation involves acquiring random samples directly from the prior, with the likelihood constraint $\mathcal{L} > 0$ not yet of consequence. We suppose that the prior is of sufficiently simple algebraic form to be sampled without much expense by using the computer's random generator. For example, the exponential prior $\text{prob}(q) = a^{-1}\exp(-q/a)$ for an intensity variable q of scale a is sampled by $q = -a \log u$, and textbooks contain a variety of transformation and related methods for obtaining samples from standard distributions.

Continuing a nested-sampling calculation involves generating a new object from the prior, subject to the likelihood constraint $\mathcal{L} > \mathcal{L}^*$. The prior domain obeying this constraint shrinks geometrically as the calculation proceeds, and we do not expect to be able to find an object in this tiny domain *ab initio*. Instead, we need to learn from experience so far, and use guidance from previous iterates. A sequence of iterates in which each uses the previous one(s) is called a Markov chain, and methods that do this in the context of random sampling are called *Markov chain Monte Carlo*, or MCMC.

In a nested-sampling iterate, there is obvious guidance built into the method itself. Whenever we need another object $\tilde{x}$ sampled from the prior $\pi(x)$ within $\mathcal{L}(x) > \mathcal{L}^*$, we already have $n-1$ such objects available as the current survivors. This strongly suggests that we should take one of the survivors (a random one in order to preserve randomness within the constraint) and copy it. Call it X:

$$\tilde{x}(\text{new}) = X(\text{old}) \qquad (\text{as introductory estimate}).$$

At least, X satisfies the desired conditions of prior and likelihood, so it's in the right general place. However, we need an independently sampled new object, not just one of the old ones over again. Hence we need to move away from X, not so far that we disobey the constraint, but far enough to lose memory of the starting point.

9.3.1 Markov chain Monte Carlo (MCMC) exploration

First, we learn how to explore the prior alone, without reference to the likelihood. After all, if we can't do the unconstrained exploration, we certainly won't be able to cope with the constraint as well. The object we seek usually has a large number of possible states $A, B, C, \ldots$, which we let the computer explore randomly across a fully-connecting pattern of permitted transitions. There can be considerable artistry involved here, using a sympathetic understanding of the structure of the problem in hand.

As might be expected, exploring a uniform prior is particularly easy. The simplest example has its states laid out linearly, with transitions allowed between neighbours.

$$A \longleftrightarrow B \longleftrightarrow C \longleftrightarrow \cdots \longleftrightarrow Z$$

Starting from the current state, choose at random to move to the left or right neighbour, and keep doing it. At the boundaries, use either wraparound conditions by linking A to Z and Z to A, or reflection conditions by linking A back to itself and Z back to itself. Similarly, in two dimensions transitions might be allowed to move randomly to any of either 4 or 8 neighbours, again with wraparound or reflective boundaries.

$$
\begin{array}{ccc}
A & \longleftrightarrow B & \longleftrightarrow C \\
\updownarrow & \updownarrow & \updownarrow \\
D & \longleftrightarrow E & \longleftrightarrow F \\
\updownarrow & \updownarrow & \updownarrow \\
G & \longleftrightarrow H & \longleftrightarrow I
\end{array}
\qquad\qquad
\begin{array}{ccc}
A & \longleftrightarrow B & \longleftrightarrow C \\
\updownarrow \times & \updownarrow \times & \updownarrow \\
D & \longleftrightarrow E & \longleftrightarrow F \\
\updownarrow \times & \updownarrow \times & \updownarrow \\
G & \longleftrightarrow H & \longleftrightarrow I
\end{array}
$$

Left for long enough, any such system will eventually be found in any specified state with uniform probability, meaning that we cannot predict where it will be unless we can model the random generator. However, in a large system with size s, it will take a long time, $\mathcal{O}(s^2)$, for knowledge of the original location to diffuse away. So, in the continuum where the number of possible states is huge, transitions are usually taken in random direction with finite distance, jumping across intermediate positions. The appropriate magnitude of this distance depends on the application, as will be seen in the 'lighthouse' example below.

The simplicity of these exploration schemes leads to a recommendation that, where possible, the original problem should be transformed into coordinates with respect to which the prior is uniform, having convenient topology such as the wraparound unit cube.

Imposing the likelihood constraint is easy. All we need to do is reject any transition that would take an object (already obeying the constraint) to a state disobeying it. Suppose, for example, that states F, H and I in the two-dimensional array above are prohibited. Transitions to them are then blocked, as shown below:

$$
\begin{array}{ccc}
A \leftrightarrow B \leftrightarrow C \\
\updownarrow \quad \updownarrow \\
D \leftrightarrow E \quad F \\
\updownarrow \\
G \quad H \quad I
\end{array}
\qquad\qquad
\begin{array}{ccc}
A \leftrightarrow B \leftrightarrow C \\
\updownarrow \,\times\, \updownarrow \,\nearrow \\
D \leftrightarrow E \quad F \\
\updownarrow \,\nearrow \\
G \quad H \quad I
\end{array}
$$

Formally, exploration continues, but with a null move if the trial transition is rejected.

9.3.2 Programming the lighthouse problem in 'C'

The aim of this application module is to solve the 'lighthouse' problem of Section 2.4, using the locations of flashes observed along the coastline to locate the lighthouse that emitted them in random directions. The lighthouse is here assumed to be somewhere in the rectangle $-2 < x < 2$, $0 < y < 2$, with uniform prior.

```
// apply.c       "LIGHTHOUSE" NESTED SAMPLING APPLICATION
//               (GNU GENERAL PUBLIC LICENSE  software: © Sivia and Skilling 2006)
//
//               u=0                                          u=1
//               ------------------------------------------
//        y=2  |::::::::::::::::::::::::::::::::::::::::::| v=1
//             |:::::::::::::::::::::::LIGHT:::::::::::::| 
//       north|::::::::::::::::::::::::HOUSE::::::::::::| 
//             |::::::::::::::::::::::::::::::::::::::::::| 
//             |::::::::::::::::::::::::::::::::::::::::::| 
//        y=0  |::::::::::::::::::::::::::::::::::::::::::| v=0
// --*------------------*----*--------*-**--**--*-*----------*---
//               x=-2              coastline -->east     x=2
// Problem:      Lighthouse at (x,y) emitted n flashes observed at D[.] on coast.
// Inputs:
//    Prior(u)   is uniform (=1) over (0,1), mapped to x = 4*u - 2; and
//    Prior(v)   is uniform (=1) over (0,1), mapped to y = 2*v; so that
//    Position   is 2-dimensional -2 <x <2 , 0 <y <2 with flat prior
//    Likelihood is L(x,y) = PRODUCT[k] (y/pi) / ((D[k] - x)^2 + y^2)
// Outputs:
//    Evidence   is Z = INTEGRAL L(x,y) Prior(x,y) dxdy
//    Posterior  is P(x,y) = L(x,y) / Z estimating lighthouse position
//    Information is H = INTEGRAL P(x,y) log(P(x,y)/Prior(x,y)) dxdy
/*_____*/

#define n    100    // # Objects
#define MAX 1000    // # iterates
/*_____*/

typedef struct
{
    double  u;        // Uniform-prior controlling parameter for x
```

```
    double   v;        // Uniform-prior controlling parameter for y
    double   x;        // Geographical easterly position of lighthouse
    double   y;        // Geographical northerly position of lighthouse
    double   logL;     // logLikelihood = ln prob(data | position)
    double   logWt;    // log(Weight), adding to SUM(Wt) = Evidence Z
} Object;
/*_____*/
double logLhood(      // logLikelihood function
double   x,           // Easterly position
double   y)           // Northerly position
{
    int   N = 64;      // # arrival positions
    double D[64] =    { 4.73,  0.45, -1.73,  1.09,  2.19,  0.12,
 1.31,   1.00,   1.32,  1.07,  0.86, -0.49, -2.59,  1.73,  2.11,
 1.61,   4.98,   1.71,  2.23,-57.20,  0.96,  1.25, -1.56,  2.45,
 1.19,   2.17,-10.66,  1.91, -4.16,  1.92,  0.10,  1.98, -2.51,
 5.55,  -0.47,   1.91,  0.95, -0.78, -0.84,  1.72, -0.01,  1.48,
 2.70,   1.21,   4.41, -4.79,  1.33,  0.81,  0.20,  1.58,  1.29,
16.19,   2.75,  -2.38, -1.79,  6.50,-18.53,  0.72,  0.94,  3.64,
 1.94,  -0.11,   1.57,  0.57};   // up to N=64 data
    int     k;                   // data index
    double logL = 0;             // logLikelihood accumulator
    for( k = 0; k < N; k++ )
        logL += log((y/3.1416) / ((D[k]-x)*(D[k]-x) + y*y));
    return logL;
}
/*_____*/
void Prior(           // Set Object according to prior
Object* Obj)          // Object being set
{
    Obj->u = UNIFORM;                    // uniform in (0,1)
    Obj->v = UNIFORM;                    // uniform in (0,1)
    Obj->x = 4.0 * Obj->u - 2.0;         // map to x
    Obj->y = 2.0 * Obj->v;               // map to y
    Obj->logL = logLhood(Obj->x, Obj->y);
}
/*_____*/
void Explore(         // Evolve object within likelihood constraint
Object* Obj,          // Object being evolved
double   logLstar)    // Likelihood constraint L > Lstar
{
    double   step = 0.1;     // Initial guess suitable step-size in (0,1)
    int      m    = 20;      // MCMC counter (pre-judged # steps)
    int      accept = 0;     // # MCMC acceptances
    int      reject = 0;     // # MCMC rejections
    Object   Try;            // Trial object
    for( ; m > 0; m-- )
    {
// Trial object
        Try.u = Obj->u + step * (2.*UNIFORM - 1.);   // |move| < step
        Try.v = Obj->v + step * (2.*UNIFORM - 1.);   // |move| < step
        Try.u -= floor(Try.u);          // wraparound to stay within (0,1)
```

```
      Try.v -= floor(Try.v);        // wraparound to stay within (0,1)
      Try.x = 4.0 * Try.u - 2.0;    // map to x
      Try.y = 2.0 * Try.v;          // map to y
      Try.logL = logLhood(Try.x, Try.y);   // trial likelihood value
// Accept if and only if within hard likelihood constraint
      if( Try.logL > logLstar )
      {  *Obj = Try;   accept++;   }
      else
          reject++;
// Refine step-size to let acceptance ratio converge around 50%
      if( accept > reject )    step *= exp(1.0 / accept);
      if( accept < reject )    step /= exp(1.0 / reject);
    }
}
/*_____*/
void Results(        // Posterior properties, here mean and stddev of x,y
Object* Samples,     // Objects defining posterior
int     nest,        // # Samples
double  logZ)        // Evidence (= total weight = SUM[Samples] Weight)
{
    double x = 0.0, xx = 0.0;     // 1st and 2nd moments of x
    double y = 0.0, yy = 0.0;     // 1st and 2nd moments of y
    double w;                     // Proportional weight
    int    i;                     // Sample counter
    for( i = 0; i < nest; i++ )
    {
        w = exp(Samples[i].logWt - logZ);
        x  += w * Samples[i].x;
        xx += w * Samples[i].x * Samples[i].x;
        y  += w * Samples[i].y;
        yy += w * Samples[i].y * Samples[i].y;
    }
    printf("mean(x) = %g, stddev(x) = %g\n", x, sqrt(xx-x*x));
    printf("mean(y) = %g, stddev(y) = %g\n", y, sqrt(yy-y*y));
}
```

The `Object` structure that encodes a possible solution needs to contain a trial location (x, y). In accordance with the recommendation to compute with a uniform prior on a unit square, x and y are slaved to controlling variables u and v, also contained in `Object` and each assigned uniform prior on $(0,1)$. The transformation in this example is simply

$$x = 4u - 2, \quad y = 2v \tag{9.18}$$

because the lighthouse is supposedly known to be within that given rectangle. However, we could equally well have encoded, say,

$$x = \tan\left[\pi\left(u - \tfrac{1}{2}\right)\right], \quad y = v/(1 - v)$$

which would have allowed the lighthouse to be located anywhere in $y > 0$ (north of the coastline), though still with a preference for $\mathcal{O}(1)$ coordinate values. Any mapping from the unit square to the upper half-plane would do. Object also contains the corresponding likelihood value $\mathcal{L}(x, y)$, as its logarithm logL for safety. The logLhood function does the numerical evaluation, with reference to the N data supplied in the vector D. Algebraically,

$$\mathcal{L}(x, y) = \prod_{k=1}^{N} \frac{y/\pi}{(D_k - x)^2 + y^2}$$

which is implicitly a function of the controlling variables u, v. Finally, Object contains the area or weight $A = h\,\mathcal{L}$, as calculated by the main program and stored in logWt.

Procedure Prior merely assigns a random (u, v) within the unit square and transforms that to (x, y) by eqn (9.18), before calculating the corresponding likelihood.

Procedure Explore takes a starting position (u, v) — in fact a copy of one of the other positions — and generates a new and supposedly independent position from it, subject to likelihood $\mathcal{L}$ exceeding the current limit $\mathcal{L}^*$. It does this by adding (or subtracting) a suitable increment to each of u and v, which are then mapped back within $(0,1)$ if they had escaped, before transforming to the desired (x, y). The increment is chosen uniformly within some range (-step, step). (Actually, a Gaussian distribution would be more usual for such purpose, but standard 'C' only provides a uniform generator.) Any trial position obeying the likelihood constraint is accepted, resulting in movement. Otherwise, the trial is rejected and there is no movement. Thus step should be reasonably large so that movement is reasonably fast, without being so large that the likelihood constraint stops all movement. It is difficult to pre-judge the appropriate size of step, which in any case tends to diminish as iterations proceed and the likelihood constraint becomes tighter. Accordingly, Explore includes a toy learning procedure which adjusts step to balance the number of accepted and rejected trials. Any such procedure in which the step-size almost certainly tends to some long-term limiting value is acceptable, and will allow the eventual position to be correctly distributed. In fact, Explore allows 20 of these MCMC steps of adjustable size, which seems to be adequate.

Finally, procedure Results computes whatever results are needed from the sequence of weighted objects found by nested sampling. The particular results chosen are the means and standard deviations of the lighthouse coordinates x and y.

9.4 Monte Carlo sampling of the posterior

At this stage, we may note that a set of 'Monte Carlo' random samples is, in fact, the preferred specification for a posterior distribution that is too complicated for brute-force enumeration or convenient analytical approximation. Alternatives are often worse. For example, it is very easy for the maximum of a multi-dimensional posterior to be way off towards one corner, and atypical of the bulk of the distribution. Similarly, the mean $\langle x \rangle$ may not be representative either, and can even be prohibited. Also it is quite hard for a distribution to stay Gaussian, because in N dimensions there are $\frac{1}{6}N^3$ third-order differentials to corrupt the $\frac{1}{2}N^2$ second-differential curvature coefficients that might define a Gaussian approximation.

True, a relatively small number of Monte Carlo samples may offer only a minuscule view of the entire space of available states. Nevertheless, they should usually suffice to quantify any desired scalar property $Q(x)$. If the n samples $\{\tilde{x}\}$ are correctly taken from the posterior $P(x)$, then their properties $\{Q(\tilde{x})\}$ will be correctly taken from the posterior distribution $P(Q)$ of that property. A dozen or more such scalar values should be sufficient to show the expected range of Q, usually as a mean and standard deviation. For example, the 12 samples $\{Q(\tilde{x})\} = \{-0.01, 0.14, 0.99, -1.50, -1.34, -0.52, -1.17,$ $0.35, -0.64, 0.80, -0.63, 0.84\}$ happened to come from the unit Gaussian distribution. While not perfect, their mean and standard deviation, $\mu \pm \sigma = -0.22 \pm 0.83$, are in fair accordance with the underlying truth, 0 ± 1. It doesn't matter that the underlying x may lie in a space of huge dimensionality, as long as we only seek low-dimensional properties Q. In such sampling, the mean of (say) 12 samples should vary around the true average by about $\pm 1/\sqrt{12} = \pm 0.3$ true standard deviations, and should deviate by more than one standard deviation only once in about 1800 trials. The expected error decreases as $\sqrt{n}$, whilst the chance of serious error decreases exponentially. Monte Carlo sampling can seem a policy of despair, but actually it works quite well.

Historically, dating back fifty years to the foundation paper of Metropolis *et al.* (1953), most algorithmic development has focussed on repeatedly sampling the posterior by some method of Monte Carlo exploration. Note, though, that posterior samples alone cannot yield the evidence integral — there's too much freedom in any plausible interpolant. Passage from (easy) sampling of the prior to (difficult) sampling of the posterior has been accomplished by methods such as simulated annealing, which might incidentally reveal the evidence as an optional by-product of the transition. In fact, the evidence has often been ignored, and to this day many authors fail to report the value of this fundamental quantity. Consequently, rational comparison of different analyses of their data is damaged. It now seems more productive to aim for the sorted likelihood function — equivalently the density of states — which underlies both evidence and posterior. As it happens, the evidence (a scalar) is a simpler quantity than the posterior (a function), which inverts the traditional view and may encourage better practice.

9.4.1 Posterior distribution

Posterior samples may be generated in proportion to the areas that contribute to Z in Fig. 9.2(b), and those are exactly the A's that we already have. The sequence of samples $\{x_k\}$, with proportional weights

$$w_k = \frac{A_k}{Z}$$

in accordance with their associated areas, models the posterior. Any property Q is available too, from the weighted sequence of values $\{Q(x_k)\}$ that defines estimates of its mean

$$\texttt{mean}(Q) = \sum_k w_k Q(x_k)$$

and standard deviation

$$\texttt{dev}(Q) = \left(\sum_k w_k \left[Q(x_k) - \texttt{mean}(Q) \right]^2 \right)^{1/2}.$$

Nested sampling usually places a useful number of samples, according to eqn (9.16) $\mathcal{O}\left(n\sqrt{\text{\# active components in data}}\right)$, across the posterior. That is usually enough to make the estimates of Q insensitive to the details of the shrinkage factors t underlying the weights w. Hence it suffices to set the factors by simply proclaiming the central values $\log(t) = -1/n$. Even so, best practice would not rely on this, but would sample the t's a dozen times or more in order to obtain corresponding estimates of a mean property, from which an average value could be derived along with its uncertainty;

$$\texttt{mean}(Q) \; = \; \text{estimate} \pm \text{numerical uncertainty}\,.$$

If this numerical uncertainty were to exceed $\texttt{dev}(Q)$, one might wish to consider re-computing with more nested samples in order to reduce the numerical uncertainty to a subsidiary level.

9.4.2 Equally-weighted posterior samples: staircase sampling

A single sample from the posterior pdf can be provided by choosing k according to $\text{prob}(k) = w_k$, and then giving the corresponding x_k. A set of at least several equally-weighted posterior samples can be used to show 'typical' $\tilde{x}$, and thus develop an intuitive understanding of the posterior. They can also be used as a compressed representation of it, stored for reference in a future when the data and such details may have been lost. Hence we now aim to draw ν samples from the sequence, ensuring a mean multiplicity $\langle n_k \rangle = \nu\, w_k$ for each object. Again, proclaiming $\log(t) = -1/n$ usually suffices to set the weights, though best practice would be to average them over a dozen or so samples of t. To minimize the reduction of information, it is desirable to keep the actual (integer) number of copies n_k as close as possible to its mean, being the integer immediately below or immediately above $\langle n_k \rangle$. Construct the cumulant staircase

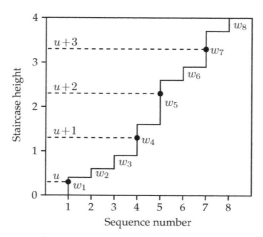

Fig. 9.6 Cumulant staircase of weights, showing extraction of four equally-weighted samples (1, 4, 5, 7) from a sequence of eight.

$$\mathcal{S}_k = u + \nu \sum_{j=1}^{k} w_j , \qquad u = \mathtt{Uniform}(0,1) ,$$

as illustrated in Fig. 9.6. Whenever the staircase $\mathcal{S}$ rises above an integer, $1, 2, 3, \ldots$, record the value of k. Each step has height νw_k, and is randomly offset vertically by u. Hence the number n_k of its intercepts has expectation $\langle n_k \rangle = \nu w_k$ as desired, and can only be either of the integers immediately above or below that value. Moreover, the total number of intercepts is exactly ν, also as sought. Of course, our number of nested samples is limited, so that these posterior samples would tend to repeat if ν were too large. To avoid this, we may restrict ourselves to

$$\nu \leqslant \left(\max_k w_k \right)^{-1} .$$

Each sample can then appear at most once in the posterior list.

9.4.3 The lighthouse posterior

Progress of the nested sampling computation of Section 9.3.2, with $n = 100$ objects, is shown in Fig. 9.7. Point k in the nested sequence is expected to be located roughly on the likelihood contour enclosing the fraction $\xi = e^{-k/n}$ of the available area. Thus the first 20 points, as shown in the first plot of Fig. 9.7, indicate the outermost $\frac{20}{100}$ of the area, $-0.2 < \log \xi < 0$. This is the domain of least likelihood, lying furthest away from the more probable locations. After 100 steps, the available area has compressed by about a factor of 'e', so that points 101–120 as shown in the second plot correspond

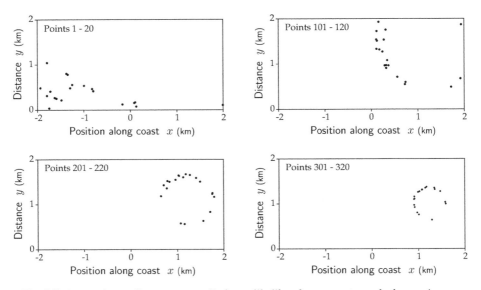

Fig. 9.7 A nested sampling sequence climbs up likelihood contours towards the maximum.

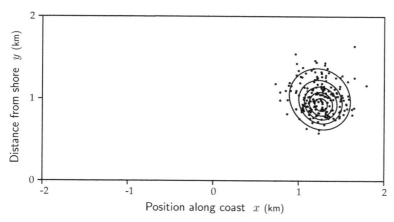

Fig. 9.8 200 posterior samples for the lighthouse problem, superposed on likelihood contours drawn at 10%, 30%, 50%, 70%, 90% of maximum.

to the shell $-1.2 < \log \xi < -1$. Points 201–220 as shown in the third plot lie in a shell $-2.2 < \log \xi < -2$ around a yet-higher likelihood contour, and so on. Eventually, the shells shrink around the point of maximum likelihood at $(1.26, 0.93)$, as the monotonic sequence of points homes in.

Nested sampling proceeds according to the shape of the likelihood contours, irrespective of the actual values. Nevertheless, it is the likelihood values that define successive objects' weights, which define the posterior and sum to Z, and which suggest when a run may be terminated. When these are used, it transpires that the evidence was $\log_e (Z/\mathrm{km}^{-64}) = -160.29 \pm 0.16$ (as the 64 data were given in kilometres), and the specified 1000 iterates were enough. Meanwhile, the position of the lighthouse as estimated from the given data was $x = 1.24 \pm 0.18$, $y = 1.00 \pm 0.19$. Figure 9.8 shows 200 posterior samples extracted from the nested sequence by staircase sampling.

9.4.4 Metropolis exploration of the posterior

As we have seen from eqns (9.15) and (9.16), nested sampling can take many iterations to reach the bulk of a confined posterior, but relatively few to cross it. Only these relatively few are significantly informative about the posterior, and are serious candidates for the equally-weighted samples that are commonly used to represent it. If the principal interest is in accurate estimation of the posterior, this imbalance can be wasteful.

Fortunately, it is almost as easy to move a sample around the posterior as it is to move it within a hard likelihood constraint. Again, we avoid complication by assuming that our transition scheme is faithful to the prior. But, instead of having a hard boundary, we now wish the chain to reach state X according to the posterior probability $P(X)$, being the already-encoded prior modulated by likelihood $\mathcal{L}(X)$.

The trick is to accept a trial move with limited probability that is proportional to $\mathcal{L}(\mathrm{destination})$. Consider transitions between two states A and B.

$$\boxed{A} \underset{\text{acceptance} \propto \mathcal{L}(A)}{\overset{\text{acceptance} \propto \mathcal{L}(B)}{\rightleftarrows}} \boxed{B}$$

These transitions are in detailed balance when A is populated proportionally to $\mathcal{L}(A)$ as desired, and B proportionally to $\mathcal{L}(B)$, because the forward and backward fluxes then balance. By analogy with physical systems, the unique stable state that ensues when every transition is in balance is called *equilibrium*. Left for long enough, a fully-connected system with detailed balance will eventually reach equilibrium and be found in state X with probability

$$\lim_{t \to \infty} \operatorname{prob}(X \,|\, \text{ignorance of generator}) = P(X).$$

All that remains is to maximize the number of acceptances in the interest of efficiency, which we do by scaling the larger of the two to the greatest possible value of 1. This gives the ansatz

$$\text{accept } (A \to B) \text{ with probability } \min\left[1, \mathcal{L}(B)/\mathcal{L}(A)\right]. \tag{9.19}$$

Even more simply, though mildly wasteful of random numbers,

$$\text{accept } (A \to B) \text{ if } \mathcal{L}(B) > \mathcal{L}(A) \times \texttt{Uniform}(0,1). \tag{9.20}$$

For comparison, nested sampling accepted $(A \to B)$ if $\mathcal{L}(B) > \mathcal{L}^*$. So, the minor extension of eqn (9.19) or (9.20) — a single extra instruction in the exploration code — allows us to explore the posterior with as many samples as we wish. Upward moves in likelihood are always accepted, but downward moves are sometimes rejected and this asymmetry gives just the required degree of modulation.

Actually, it is not strictly necessary to seed the exploratory Markov chains with pre-computed posterior samples. Any seed would suffice in principle, and the chain would eventually lose memory of it (though this might take a long time if the seed was highly improbable). Historically, this is how the whole subject of Bayesian computation started fifty years ago, with Metropolis *et al.* (1953).

9.5 How many objects are needed?

As we investigate the number of objects needed, an unexpected advantage of nested sampling comes to light. Geometrically, the permitted domain shrinks by about 1 part in n per iterate. This concentrates the n objects ever more tightly into regions of higher likelihood, as illustrated in Fig. 9.9. At each iterate, the worst (outermost) object is discarded in favour of a copy of an internal survivor, which is then re-equilibrated.

9.5.1 Bi-modal likelihood with a single 'gate'

The likelihood function need not have the convenient single maximum shown in Fig. 9.9. Consider instead a bi-modal likelihood, having two maxima, such as that shown in Fig. 9.10. One mode is *dominant* because it contains the bulk of the evidence $\int \mathcal{L} \, d\xi$;

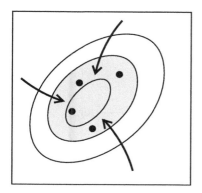
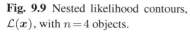

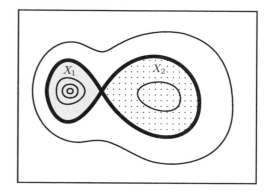

Fig. 9.9 Nested likelihood contours, $\mathcal{L}(\boldsymbol{x})$, with $n=4$ objects.

Fig. 9.10 The thick likelihood contour around the dominant mode X_1 is a 'gate'. The subordinate mode X_2 is surrounded by a 'trap'.

the other is *subordinate*. There is a critical likelihood *gate* below which the modes are connected, and above which they are separate. Before the gate is reached, MCMC exploration can presumably diffuse freely around the volume enclosing both modes. After the gate is passed, transitions between modes need to jump across to the other tiny but relatively distant domain, so are essentially blocked and an object can diffuse only within its own mode.

At the critical likelihood, let the accessible volumes be X_1 for the dominant mode and X_2 for the subordinate. The chance of an exploratory object falling into the dominant mode as the gate closes behind it is the proportional gate width

$$W = \frac{X_1}{X_1 + X_2} .$$

Conversely, with chance $1-W$, it falls into the subordinate mode, where it is essentially trapped. With n independent objects, the chance of 'success' with at least one object in the dominant mode is

$$\text{prob(success}|n \text{ objects}) = 1 - (1-W)^n .$$

Basically, we need rather more than W^{-1} objects to be reasonably sure of at least one success. Thus, if the gate width is $W = 1/64$ but we only supply $n = 10$ objects, then the chance of a success is less than $1/6$.

9.5.2 Multi-modal likelihoods with several 'gates'

It may be that a particular multi-modal problem has just one narrow gate. Eventually, the likelihood will favour the dominant mode by a factor that more than compensates for the narrow opening, but that's not known as the gate is passed.

It is perhaps more likely that a complicated problem has several gates, perhaps six gates of width $1/2$ or so at different likelihood levels, instead of just one. After all, there

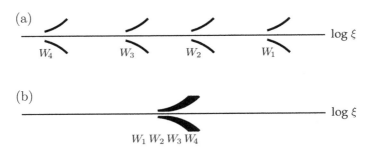

Fig. 9.11 (a) Several wide gates, and (b) one narrow gate of equivalent aperture.

is a bigger parameter-space associated with this more general framework. If objects are programmed to explore independently, half will fail at the first gate, then half the survivors will fail at the second and so on until the final survival rate is only $1/64$, just as for a single narrow gate; this is shown schematically in Fig. 9.11.

At each iterate of nested sampling, though, the object with lowest (worst) likelihood is eliminated in favour of a copy of one of the (better) others. After a gate is passed and the likelihood constraint continues to climb to more-restrictive heights, the subordinate mode should become progressively less populated. Indeed, after the constraint has climbed above the subordinate maximum, that mode can have no surviving objects at all. So, even if a gate is quite narrow, the dominant mode becomes re-populated provided at least one object manages to find it.

With a gate width of $1/2$, the chance of having at least one success from n objects is $1 - 2^{-n}$, nearly certain. Before the next gate is reached, it may well be that the population along the dominant route has increased from the original $n/2$ back up to or nearly to n. After 6 such gates, with a mere 2^{-n} chance of failure at each, the chance of success is $(1 - 2^{-n})^6$. Not only is this more than 99% for $n = 10$, but quite soon afterwards most or all of the objects should be in the dominant mode. Nested sampling's 'copy' operation has turned an expectation of total failure into a a high probability of complete success.

Generally, with a series of well-separated gates of widths W_g, the expectations of success are

$$\text{prob(success}|n\text{ objects)} = \begin{cases} 1 - \left(1 - \prod_g W_g\right)^n & \text{for no copying,} \\ \prod_g \left[1 - (1 - W_g)^n\right] & \text{with copying.} \end{cases}$$

Basically, the number of objects needed to give a good chance of success is

$$n_{\min} \approx \begin{cases} \prod_g \left[W_g^{-1}\right] & \text{for no copying,} \\ (W_{\max})^{-1} & \text{with copying.} \end{cases}$$

and copying always beats exploration by individually-preserved objects.

9.6 Simulated annealing

Nested sampling can be compared with the traditional method of *simulated annealing*, which uses fractional powers $\mathcal{L}^\beta$ of the likelihood to move gradually from the prior ($\beta = 0$) to the posterior ($\beta = 1$). As the 'coolness' β increases, annealing gently compresses a set of points $\tilde{x}$ sampled from $\mathrm{d}P_\beta \propto \mathcal{L}^\beta \mathrm{d}\xi$, known as a thermal *ensemble*. At stage β, the mean log-likelihood

$$\langle \log \mathcal{L} \rangle_\beta = \int \log \mathcal{L} \, \mathrm{d}P_\beta = \frac{\int \mathcal{L}^\beta \log \mathcal{L} \, \mathrm{d}\xi}{\int \mathcal{L}^\beta \, \mathrm{d}\xi} = \frac{\mathrm{d}}{\mathrm{d}\beta}\left[\log\left(\int \mathcal{L}^\beta \, \mathrm{d}\xi\right)\right]$$

is estimated by averaging over the corresponding ensemble. Summing this yields

$$\int_0^1 \langle \log \mathcal{L} \rangle_\beta \, \mathrm{d}\beta = \log\left(\int \mathcal{L} \, \mathrm{d}\xi\right) - \log\left(\int \mathrm{d}\xi\right) = \log Z$$

which is the *thermodynamic integration formula* for the evidence Z. The bulk of the ensemble, with respect to $\log \xi$, should follow the posterior $\mathrm{d}P_\beta \propto \mathcal{L}^\beta \xi \, \mathrm{d}\log \xi$ and be found around the maximum of $\mathcal{L}^\beta \xi$. Under the usual conditions of differentiability and concavity '⌢', this maximum occurs where

$$\frac{\mathrm{d}\log \mathcal{L}}{\mathrm{d}\log \xi} = -\frac{1}{\beta} .$$

Annealing over β thus tracks the $\log \mathcal{L} / \log \xi$ slope, whereas nested sampling tracks the underlying abscissa value $\log \xi$.

9.6.1 The problem of phase changes

As β increases from 0 to 1, one hopes that the annealing maximum tracks steadily up in $\mathcal{L}$, so inward in ξ; this is shown schematically in Fig. 9.12(a). The *annealing schedule*

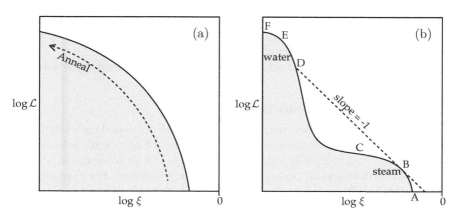

Fig. 9.12 Proper annealing needs log-likelihood to be concave like (a), not (b).

that dictates how fast β is increased ought to allow successive posteriors P_β to overlap substantially — exactly how much is still a matter of some controversy. Yet it may not be possible at all.

Suppose that $\mathcal{L}^\beta \xi$ is not concave, as in Fig. 9.12(b). No matter what schedule is adopted, annealing is supposed to follow the concave hull of the log-likelihood function as its tangential slope flattens. But this will require jumping right across any convex '$\smile$' region that separates ordinary concave 'phases' where local maxima of $\mathcal{L}^\beta \xi$ are to be found. At $\beta = 1$, the bulk of the posterior should lie near a maximum of $\mathcal{L}\xi$, in one or other of these phases. Let us call the outer phase 'steam' and the inner phase 'water', as suggested by the potentially large difference in volume. Annealing to $\beta = 1$ will normally take the ensemble from the neighbourhood of A to the neighbourhood of B, where the slope is $\mathrm{d}\log\mathcal{L}/\mathrm{d}\log\xi = -1/\beta = -1$. Yet we actually want samples to be found from the inner phase beyond D, finding which will be exponentially improbable unless the intervening convex valley is shallow. Alternatively, annealing could be taken beyond $\beta = 1$ until, when the ensemble is near the point of inflection C, the supercooled steam crashes inward to chilled water, somewhere near F. It might then be possible to anneal back out to unit temperature, reaching the desired water phase near E. However, annealing no longer bridges smoothly during the crash, and the value of the evidence is lost. Along with it is lost the internal Bayes factor $\mathrm{prob}(\text{states near E})/\mathrm{prob}(\text{states near B})$ which might have enabled the program to assess the relative importance of water and steam. If there were three phases instead of just two, annealing might fail even more spectacularly. It would be quite possible for supercooled steam to condense directly to cold ice, and superheated ice to sublime directly to hot steam, without settling in an intermediate water phase at all. The dominant phase could be lost in the hysteresis, and inaccessible to annealing.

Phase change problems in general are well known to be difficult to anneal. Nested sampling, though, marches steadily down in prior mass ξ along ABCDEF$\cdots$, regardless of whether the associated log-likelihood is concave or convex or even differentiable at all. There is no analogue of temperature, so there is never any thermal catastrophe. Nested points will pass through the steam phase to the supercooled region, then steadily into superheated water until the ordinary water phase is reached, traversed and left behind in an optional continued search for ice. All the internal Bayes factors are available, so the dominant phase can be identified and quantified.

This, then, is a second illustration of the ability of nested sampling to solve multi-phase problems.

9.6.2   Example: order/disorder in a pseudo-crystal

Consider the following elementary model of *order/disorder* based on 'switches', each of which can be in either of two states, 0 or 1. A sequence of M switches is laid out along a line, so there are 2^M equally-weighted prior states. The switches define a sequence of clusters c with widths h_c across which the state is constant. For example, the ten switches 0001111001 have four clusters, respectively three 0's (width $h_1 = 3$), followed by four 1's ($h_2 = 4$), then two 0's ($h_3 = 2$) and finally a single 1 ($h_4 = 1$). Each cluster has an energy benefit (i.e. a log-likelihood gain) proportional to the number $\frac{1}{2}h(h-1)$

of internal interactions permitted among its members, so that (with specific scaling)

$$\log \mathcal{L} = \frac{2}{M} \sum_c \tfrac{1}{2} h_c (h_c - 1), \qquad \sum_c h_c = M.$$

The example state 0001111001 has $\log \mathcal{L} = 2$. Of the 1024 states of 10 switches, the top two (0000000000 and 1111111111) are fully ordered with $\log \mathcal{L} = 9$ and share 49% of the posterior, the next four (0000000001, 0111111111, 1000000000, 1111111110) with $\log \mathcal{L} = 7.2$ share another 16%, and so on down to the two lowest states (0101010101 and 1010101010) with $\log \mathcal{L} = 0$ which share 0.006%.

Figure 9.13 shows the behaviour for $M = 1000$, precisely calculated by recurrence on M. An 'order' phase with wide clusters dominates, having the fully-ordered states $0000 \cdots$ and $1111 \cdots$ with $\log \mathcal{L} = 999$ sharing 71% of the posterior, the next four with $\log \mathcal{L} = 997.002$ sharing 19% and so on. With M being quite large, there is also a well-separated 'disorder' phase with most clusters narrow, and the 'order' phase is favoured overall by a Bayes factor of e^{300}.

An ensemble annealed to $\beta = 1$, though, has no chance (technically, about e^{-110} chance) of finding the tiny volume occupied by the 'order' states. It ought to transition to the ordered phase at the freezing point $\beta = 0.69$ where the two phases ought to become equally populated, but it won't. As expected, the simulation in Skilling (2004), which started with a random state and evolved by inverting atoms at random according to

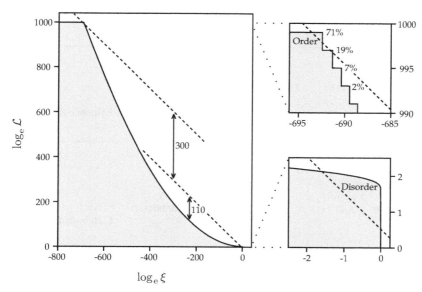

Fig. 9.13 Order/disorder example for 1000 switches. The upper sub-plot magnifies the 'order' phase, and the lower sub-plot magnifies the 'disorder' phase. The order phase is favoured by a Bayes factor exp(300) but is hard to find by a factor exp(110). Dashed lines enclose 75% of posterior samples for each phase.

the usual detailed balance, failed to move away from the 'disorder' phase in an allotted trillion trial inversions. Even if by incredible luck it had succeeded in finding the 'order' phase, it could not have determined the evidence $Z = e^{307}$.

Yet, nested sampling (with a fresh sample within the likelihood constraint approximated by allowing trial MCMC inversions of each switch ten times per iterate, and a collection of any size n) successfully estimates $\log \mathcal{L}$ as a function of $\log \xi$ and reaches the fully-ordered states steadily, in the expected $n\mathcal{H} \approx 700\,n$ iterates.

9.6.3 Programming the pseudo-crystal in 'C'

```
// apply.c   "PSEUDO-CRYSTAL" NESTED SAMPLING APPLICATION
//           (GNU GENERAL PUBLIC LICENSE software: © Sivia and Skilling 2006)

// Problem:       M switches s = 0 or 1, grouped in clusters of widths h.
//                   e.g. M=10, s = {0,0,0,1,1,1,1,0,0,1}
//                                h = { 3  ,  4   , 2 ,1}
// Inputs:
//    Prior(s)    is uniform, 1/2^M on each of 2^M states
//    Likelihood  is L(s) = exp( SUM h(h-1)/M )
// Outputs:
//    Evidence    is Z = SUM L(s) Prior(s)
//    Posterior   is P(s) = L(s) Prior(s) / Z
//    Information  is H = SUM P(s) log(P(s)/Prior(s))
/*_____*/
#define n    1      // # Objects
#define MAX 800     // # iterates
#define M  1000     // # switches in this application
/*_____*/
typedef struct
{
    char   s[M];   // state of switches
    double logL;   // logLikelihood = ln prob(data | s)
    double logWt;  // log(Weight), adding to SUM(Wt) = Evidence Z
} Object;
/*_____*/
double logLhood(   // logLikelihood function
char*  s)          // switches
{
    int    i, j;        // left and right counters
    double logL = 0;    // logLikelihood accumulator
    i = 0;                                        // L.H. boundary
    for( j = 1; j < M; j++ )
        if( s[j] != s[j-1] )
            {                                     // R.H. boundary found
                logL += (j - i) * (j - i - 1);    // cluster width h = j-i
                i = j;                            // reset L.H. boundary
            }
    logL += (j - i) * (j - i - 1);                // R.H. cluster
    return logL / M + sqrt(DBL_EPSILON) * UNIFORM;   // normalized
}                                   // jitter eliminates ties between likelihood values
/*_____*/
```

```
void Prior(              //  Set Object according to prior
Object* Obj)             //  Object being set
{
    int   j;
    for( j = 0; j < M; j++ )
        Obj->s[j] = (int)(2 * UNIFORM) % 2; //  0 or 1
    Obj->logL = logLhood(Obj->s);
}
/*_____*/
void Explore(            //  Evolve object within likelihood constraint
Object* Obj,             //  Object being evolved
double  logLstar)        //  Likelihood constraint L > Lstar
{
    int       m = 10 * M;      //  MCMC counter (pre-judged # steps)
    int       try;             //  Try flipping this switch
    double    logLtry;         //  Trial loglikelihood
    for( ; m > 0; m-- )
    {
        try = (int)(M * UNIFORM) % M;      //  random switch
        Obj->s[try] = 1 - Obj->s[try];     //  try flipping
        logLtry = logLhood(Obj->s);        //  trial loglikelihood
        if( logLtry > logLstar )
            Obj->logL = logLtry;           //  accept
        else
            Obj->s[try] = 1 - Obj->s[try]; //  reject
    }
}
/*_____*/
void Results(            //  Output nested sampling sequence
Object* Samples,         //  Objects defining posterior
int     nest,            //  # Samples
double  logZ)            //  Evidence (= total weight = SUM[Samples] Weight)
{
    int   k;             //  Sample counter
    for( k = 0; k < nest; k++ )
        printf("%7.2f %8.4f\n", -(k+1.) / n, Samples[k].logL);
                         //  print log(enclosed prior mass), log(likelihood)
}
```

The aim of the above module is to estimate the density of states by plotting $\log \mathcal{L}$ as a function of $\log \xi$, in order to recover the theoretical Fig. 9.13. Hence the Object structure needs to contain a vector s of $M = 1000$ switches, as well as the usual log-likelihood and logarithmic weight as used by the main program.

Procedure Prior just sets the switches at random, before calculating the corresponding likelihood. This being a discrete problem, a small amount of random jitter is added to each likelihood, in the logLhood function. After all, nested sampling is trying to compress the available domain by a factor of $\exp(1/n)$ or so each step, and this would be difficult if a single likelihood value occupied more prior mass than that.

Procedure Explore takes a starting state s — in fact a copy of one of the other positions — and generates a new and supposedly independent state from it, subject to likelihood $\mathcal{L}$ exceeding the current limit $\mathcal{L}^*$. It does this by trying to flip each of the

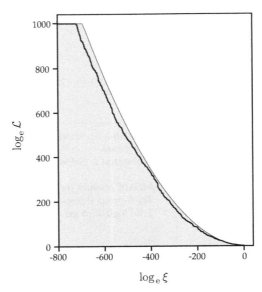

Fig. 9.14 Nested sampling (line) recovers the theoretical shape (shaded) of Fig. 9.13.

1000 switches about 10 times. Any trial position obeying the likelihood constraint is accepted, resulting in movement. Otherwise, the trial is rejected and there is no change. It would be more efficient to calculate only the local changes induced by flipping a switch, but for transparency the likelihood is here computed from scratch.

Finally, procedure `Results` prints out the nested sampling trajectory, ready for plotting as Fig. 9.14.

Even though only one object was used in this run, the density curve is reproduced well, with just the expected random drifts reflected in an evidence estimate $\log_e Z = 280 \pm 27$ lying a plausible one standard deviation from the true value of 307. The two-phase character of the system shows up clearly, as does its convex nature indicating the difficulty of annealing it. Nested sampling took 720 iterates to reach the fully-ordered states, consistent with the expected $\mathcal{H} \pm \sqrt{\mathcal{H}}$ with $\mathcal{H} = 692$.

10 Quantification

In this chapter, we extend the basic exploratory techniques of Chapter 9, aiming to answer the perennial questions *'What's there, and how much?'*. As so often in algorithm design, there are some nice tricks to be found. Here are some of them.

10.1 Exploring an intrinsically non-uniform prior

We now consider problems in which the discrete states of a system are assigned different prior probabilities, which cannot usefully be transformed to uniformity. It is possible to choose potential transitions at random, as before. Indeed, many writers recommend just that, using rejections to control the desired non-uniformity, as in Metropolis exploration described in Section 9.4.4. Our view, though, is that most priors are sufficiently simple to be explored efficiently without rejection, and that the insight required to accomplish this is worthwhile.

Each transition, $A \rightarrow B$ and so on, is assigned a corresponding rate r_{BA}. If the system starts in state A it will evolve to B at rate r_{BA} per unit time:

$$\text{prob}(B|A) = r_{BA}\,dt \qquad (10.1)$$

in a small interval dt. Then, if our knowledge of the system is uncertain, with probability $\pi(A)$ of being in A, the flux from A to B in time dt will be $\pi(A)\,r_{BA}\,dt$. Practical schemes adopt the *detailed balance* convention

$$\boxed{A} \underset{r_{AB}}{\overset{r_{BA}}{\rightleftarrows}} \boxed{B} \qquad \text{with} \quad \frac{r_{BA}}{r_{AB}} = \frac{\pi(B)}{\pi(A)} \qquad (10.2)$$

in which forward and backward fluxes become equal when the states are populated according to their priors.

From current state X, the total rate to other states is $R = \sum_{Y \neq X} r_{YX}$, whence the interval τ to the next transition away from X will be distributed as

$$\text{prob}(\tau) = R \exp(-R\tau). \qquad (10.3)$$

So we can jump directly to the next transition by sampling

$$\tau = -R^{-1} \log\left[\texttt{Uniform}(0,1)\right]. \qquad (10.4)$$

When a transition falls due, we list the rates r_{YX} for all available destinations Y and select Y according to

$$\text{prob}(Y|X) = \frac{r_{YX}}{R}. \qquad (10.5)$$

Finally, when the accumulated time would cross an assumed *equilibration time*, we deem the current state to be the (sufficiently) independent new sample that we seek. In this way, the prior is explored without any wasteful rejection.

Incorporating the likelihood constraint is easy. All we do is reject any transition that would lead to a destination disobeying the constraint. Time is incremented as before, but the source state is preserved. As before, the prior is explored freely for transitions within the constraint, but transitions to outside are forbidden.

10.1.1 Binary trees for controlling MCMC transitions

As a technical trick, identifying the next transition from a choice of M can be done in $\log(M)$ operations instead of the more obvious M, provided the transition scheme is created with this in mind. Pad the M rates (say 5 of them, r_1, r_2, r_3, r_4, r_5) with zeros until the vector length M^+ is a power of 2 (here $M^+ = 2^3 = 8$). Then arrange a binary tree of partial sums.

$r_1+r_2+r_3+r_4+r_5$							
$r_1+r_2+r_3+r_4$				r_5			
r_1+r_2		r_3+r_4		r_5		0	
r_1	r_2	r_3	r_4	r_5	0	0	0

This involves doubling the power-of-2 storage but requires only a single $\mathcal{O}(M)$ pass to set up. Thereafter, if a local transition from X to a nearby Y only affects one or a few of the base rates, the tree can be kept up-to-date by updating only those few base cells and the $\log_2 M^+$ cells above each. Selecting a particular rate involves finding the base cell j at which the cumulant

$$r_1 + r_2 + \ldots + r_j > fR, \quad R = r_1 + r_2 + \ldots + r_m \tag{10.6}$$

first exceeds some random fraction f of the total rate R. With the tree in place, this can easily be done in $\log_2 M^+$ steps by using the partial sums to switch left or right at each descending level. As a by-product, the total R is always available at the top of the tree.

Addressing within a binary tree is particularly slick if storage addresses are assigned as follows, with the total at address #1.

1							
2				3			
4		5		6		7	
8	9	10	11	12	13	14	15

The otherwise-unused address #0 holds M^+, making the tree self-contained, of total size $2 \times M^+$.

The accompanying procedures `PutRate` and `GetRate` update and use the trees. `PutRate` encodes the obvious pattern of additions consequent upon inserting a new individual rate value. `GetRate` selects a random rate from the total, which involves subtraction as the tree is descended from its apex to identify the component covering the randomly-selected cumulant as in eqn (10.6). In floating-point arithmetic, subtraction amplifies relative rounding errors, leading to the possibility of selecting a component known to have zero rate. Such anomalous behaviour is avoided by GetRate, the proof of robustness being in the accompanying comments.

```
// tree.c              PROCEDURES FOR BINARY TREE
//            (GNU GENERAL PUBLIC LICENSE software: © Sivia and Skilling 2006)

// Example:  m=5 rates (r0,r1,r2,r3,r4) are stored using
//           mplus=8 cells, and are preceded by partial sums.
//   +------------------------------------------------------------+
// T |1                    r0+r1+r2+r3+r4                          |
//   |------------------------------------------------------------|
// R |2          r0+r1+r2+r3          |3              r4           |
//   |-------------------------------|----------------------------|
// E |4   r0+r1    |5    r2+r3    |6      r4      |7       0        |
//   |-------------|--------------|--------------|----------------|
// E |8   r0 |9  r1 |10 r2 |11 r3 |12 r4 |13  0 |14  0 |15   0 |
//   +------+------+------+------+------+------+------+------+
//           Tree[0] holds the value of mplus, here 8.
/*_____*/
void PutRate(      // Update binary tree with new rate
double* Tree,      // Binary tree of rates
int     j,         // Input cell (not including mplus)
double  r)         // New value of rate
{
    j += (int)Tree[0];
    Tree[j] = r;
    for( ; j > 1; j >>= 1 )
        Tree[j>>1] = Tree[j] + Tree[j^1];
}
/*_____*/
int GetRate(       // Select cell (not including mplus) with random rate
double* Tree)      // Binary tree of rates
{   // This procedure is crafted never to select a cell having rate=0
    double f;
    int    j;
    int    mplus = (int)Tree[0]; // Assuming Tree[1] > 0, then
    f = Tree[1] * UNIFORM;        // 0 < f ≤ Tree[1] for any rounding.
    for( j = 1; j < mplus; )
    {                  // Enter with
        j <<= 1;       // f ≤ Tree[parent] = Tree[j]+Tree[j+1] ≥ Tree[j]
        if( f > Tree[j] ) // Only possible if Tree[j] < Tree[parent],
        {                 // so that Tree[j+1] > 0,
            f -= Tree[j++]; // hence j+1 is a safe destination.
            if( f > Tree[j] ) // Rounding error occasionally
                f = Tree[j];  // matters, so keep f ≤ Tree[j].
        }                 // Either way, f > 0 by construction.
    }                  // Else 0 < f ≤ Tree[j], so Tree[j] > 0,
    return j - mplus;  // hence the entry j was a safe destination.
}
```

10.2   Example: ON/OFF switching

We now specialize to problems in which the object in question is a set of N binary ON/OFF switches. This model is the kernel of a variety of problems involving mixtures or combinations of constituents. Each switch can be ON (1) or OFF (0) with pre-assigned prior probability

$$\text{prob}\left(\text{switch } s \text{ is } \left\{ \begin{matrix} \text{ON} \\ \text{OFF} \end{matrix} \right\} \right) = \left\{ \begin{matrix} \pi_s \\ 1 - \pi_s \end{matrix} \right\}.$$

There are 2^N states in all, and we need a transition scheme that reaches all of them.

10.2.1   The master engine: flipping switches individually

The simplest, most obvious, scheme allows one switch at a time to flip ON or OFF. This is the master transition scheme, known as an *engine*, but other engines may also be useful. Clearly all states are accessible, but not always in a single step. From any particular state, any single switch can be flipped, yielding N transitions. In accordance with detailed balance, switching ON from OFF should have a rate proportional to π, whilst OFF from ON should have $1 - \pi$ instead.

For example, the state 00110 in which the third and fourth switches out of five are ON would have available transitions

$$r = \kappa \left(\pi_1, \pi_2, 1 - \pi_3, 1 - \pi_4, \pi_5 \right)$$

to states 10110, 01110, 00010, 00100, 00111 respectively, where κ is some normalizing engine *throttle* setting. κ is also the rate at which each switch equilibrates. The natural setting is $\kappa = 1$, for which equilibration times are $\mathcal{O}(1)$. We do not attempt to maximize the individual rates, Metropolis-style, because that would make some switches equilibrate slower than others. Finding the interval τ to the next transition involves the total rate R, where

$$R = \kappa \left[\pi_1 + \pi_2 + (1 - \pi_3) + (1 - \pi_4) + \pi_5 \right].$$

Selecting this next transition involves sampling from r/R. Performing it involves flipping the switch ON/OFF and inverting its rate ($\pi \longleftrightarrow 1 - \pi$). All the book-keeping can be done at cost $\log N$ by using a binary tree of the N rates available at the current time. All transitions that fall within the likelihood constraint are accepted, because the rates have been set to explore the prior faithfully.

10.2.2   Programming which components are present

The above ideas are coded into the following 'C' application module for estimating which components of a mixture are present. In this little problem, there may be five components, coded as M = 5 switches s, each of which has its prior probability π, coded as PrON, of being ON. The likelihood function has three factors, the first of which claims it unlikely (prob = 0.2) that either component 2 or component 4 is ON. On the other hand, the second factor claims that either 3 or 4 is rather likely to be ON, with probability 0.9. The third factor claims that the probability of both 1 and 4 being ON is 0.8. Such data idealize the sort of information often available in spectrometry, where individual lines relate ambiguously to components of a mixture.

In the code, the Object structure holds the 5 switch settings s, as well as their log-likelihood logL and the weight logWt controlled by the main program. It also holds the corresponding binary tree of weights, dimensioned through Mplus, being the power-of-2 next above M (i.e. 8).

```
// apply.c   "ON/OFF SWITCHING" NESTED SAMPLING APPLICATION
//           (GNU GENERAL PUBLIC LICENSE software: (c) Sivia and Skilling 2006)

// Problem:       5 switches s, with probabilistic data.
// Inputs:
//    Prior(s)    is {0.1, 0.2, 0.3, 0.2, 0.1} for the 5 switches.
//    Likelihood  is Pr(s[2] OR s[4]) = 0.2, Pr(s[3] OR s[4]) = 0.9,
//                   Pr(s[1] AND s[4]) = 0.8.
// Outputs:
//    Evidence    is Z = SUM L(s) Prior(s)
//    Posterior   is P(s) = L(s) Prior(s) / Z
//    Information  is H = SUM P(s) log(P(s)/Prior(s))
/*_____*/

#define n     100      // # Objects
#define MAX   1000     // # iterates
#define M     5        // # switches
#define Mplus 8        // power-of-2 >= M
#include "tree.c"  // binary tree procedures void PutRate, int GetRate
const double PrON[M] = {0.1, 0.2, 0.3, 0.2, 0.1};   // Prior(s is ON)
/*_____*/

typedef struct
{
    char    s[M];            // switches
    double  Tree[2*Mplus];   // binary rate tree
    double  logL;            // logLikelihood = ln prob(data | s)
    double  logWt;           // log(Weight), with SUM(Wt) = Evidence Z
} Object;
/*_____*/

double logLhood(          // logLikelihood function
char*   s)                // switches
{
    double  L = 1.0;
    L *= (s[2] | s[4]) ? 0.2 : 0.8;
    L *= (s[3] | s[4]) ? 0.9 : 0.1;
    L *= (s[1] & s[4]) ? 0.8 : 0.2;
    return  log(L) + sqrt(DBL_EPSILON) * UNIFORM;
}                         // jitter eliminates ties between likelihood values
/*_____*/
```

```
void Prior(                    // Set Object according to prior
Object* Obj)                   // Object being set
{
    int  i;
// Initialize empty Tree of transition rates
    Obj->Tree[0] = Mplus;
    for( i = 1; i < 2*Mplus; i++ )
        Obj->Tree[i] = 0.0;
// Initialize object
    for( i = 0; i < M; i++ )
    {
        Obj->s[i] = (UNIFORM < PrON[i]);
        PutRate(Obj->Tree, i, Obj->s[i] ? 1.-PrON[i] : PrON[i]);
    }
    Obj->logL = logLhood(Obj->s);
}
/*_____*/
void Explore(                  // Evolve object within likelihood constraint
Object* Obj,                   // Object being evolved
double  logLstar)              // Likelihood constraint L > Lstar
{
    double Interval = 30.0; // pre-judged time to equilibrate
    double t = 0.0;            // evolution time, initialized 0
    double logLtry;           // trial logLikelihood
    int    i;                 // switch being flipped
    while( (t += -log(UNIFORM) / Obj->Tree[i]) < Interval )
    {
        i = GetRate(Obj->Tree);
        Obj->s[i] = 1 - Obj->s[i];             // trial state
        logLtry = logLhood(Obj->s);
        if( logLtry > logLstar )
        {                                      // accept
            Obj->logL = logLtry;
            PutRate(Obj->Tree, i,
                    Obj->s[i] ? 1.-PrON[i] : PrON[i]);
        }
        else
            Obj->s[i] = 1 - Obj->s[i];         // reject
    }
}
/*_____*/
void Results(        // Posterior properties, here mean of s
Object* Samples,     // Objects defining posterior
int     nest,        // # Samples
double  logZ)        // Evidence (= total weight = SUM[Samples] Weight)
{
    double post[M] = {0,0,0,0,0};   // posterior prob(ON)
    double w;                        // Proportional weight
    int    i;                        // Sequence counter
    int    k;                        // Sample counter
    for( i = 0; i < nest; i++ )
```

```
{
    w = exp(Samples[i].logWt - logZ);
    for( k = 0; k < M; k++ )
        if( Samples[i].s[k] )  post[k] += w;
}
for( k = 0; k < M; k++ )
    printf("%d:  Posterior(ON)=%3.0f%%\n", k, 100.*post[k]);
}
```

Procedure `Prior` sets the switches at random, according to their prior, before calculating the corresponding likelihood — including the small random jitter recommended for a discrete problem. It also initializes the binary tree of transition rates through `PutRate`. Procedure `Explore` sets up a time `Interval` after which the evolving sample is deemed to be sufficiently independent of its starting state. Meanwhile, the clock-time `t` accumulates the sub-intervals between transitions as calculated by eqn (10.4). At each transition moment, the appropriate switch is selected by `GetRate` according to eqn (10.5), and flipped. Then the likelihood constraint is applied to determine if the transition is to be accepted, in which case the new likelihood is adopted, or whether it is to be rejected, in which case the flipped switch is reset to its previous state. Finally, the `Results` procedure accumulates the posterior probability of each switch being ON.

This being such a small problem, correct results can be evaluated by brute force addition over the 32 possible switch settings, and compared with the computed results

	Correct	Computed
$\log_e Z$	-3.36	-3.32^*
$\mathcal{H}$	0.63	0.61
prob(s_0 is ON)	10%	6%
prob(s_1 is ON)	25%	25%
prob(s_2 is ON)	13%	14%
prob(s_3 is ON)	61%	63%
prob(s_4 is ON)	17%	14%

* with ± 0.08 numerical uncertainty

There is no significant or systematic difference between correct and computed results.

10.2.3 Another engine: exchanging neighbouring switches

Neighbouring components in a mixture commonly have substantial similarities, so that their switches might be correlated. For efficient exploration, there could be explicit transitions between switches that are known to be related. For example, an ON setting could move directly between the fourth and fifth positions

$$\cdots 10 \cdots\cdots \quad \longleftrightarrow \quad \cdots 01 \cdots\cdots$$

instead of having to move inefficiently in two steps via $\cdots 00 \cdots\cdots$ or $\cdots 11 \cdots\cdots$ (and hoping that at least one of these intermediate states was consistent with the data, so would not be prohibited).

To accord with detailed balance, the transition rate from the former state to the latter should be

$$r_{...01......\,,\,...10......} = \kappa'\,(1-\pi_4)\,\pi_5\,. \tag{10.7}$$

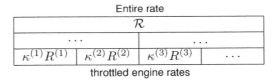

This second transition engine has its own throttle setting κ', which may differ from the first engine's throttle. Indeed, there could be several such engines, each designed to pick out transitions that are sympathetic to the particular application in hand.

10.2.4   The control of multiple engines

If there are several engines, each of them $1, 2, 3, \ldots$ has its family of transition rates, culminating in the total rate $R^{(1)}, R^{(2)}, R^{(3)}, \ldots$ at which it is supplying events. The entire rate at which all the engines together are working is $\mathcal{R}$, where

$$\mathcal{R} = \kappa^{(1)}R^{(1)} + \kappa^{(2)}R^{(2)} + \kappa^{(3)}R^{(3)} + \cdots$$

As above, finding the interval τ to the next transition involves this entire rate, and selecting the appropriate engine involves sampling from $\kappa\,R/\mathcal{R}$, after which its appropriate transition is selected.

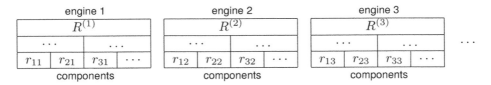

Generally, it is recommended practice to tune the throttle settings κ in order to keep the computational resources allotted to different engines more-or-less balanced. Then, no individual engine will dominate the CPU, and even if one of them turns out to be less efficient for the particular problem in hand, the waste of resources will not be overwhelming.

10.3   Estimating quantities

Commonly in applications, each ON switch is accompanied by an associated positive quantity q, OFF being indistinguishable from $q = 0$. In spectrometry for example, q might be the concentration of a component chemical that may or may not be present in the source. If it is present, with prior probability assigned as π, it would have an associated amount q, specified perhaps by an exponential prior or (often better)

$$\text{prob}(q) = a/(a+q)^2 \qquad (10.8)$$

which expresses a preference for q being $O(a)$, whilst not discriminating strongly against occasional larger values. Presence and quantity could be combined into a single prior

$$\text{prob}(q) = (1-\pi)\delta(q) + \pi a/(a+q)^2$$

but it's computationally more efficient to keep them separate. In accordance with the recommendation to have such quantities slaved to controlling variables with respect to which the prior is uniform, we let

$$q = a\,\frac{u}{1-u}\,, \quad u = \frac{q}{a+q}\,, \qquad (10.9)$$

where q is controlled by u having flat prior over $(0,1)$. Specifically, we sample q from its prior by sampling

$$u = \texttt{Uniform}(0,1) \qquad (10.10)$$

and substituting back.

We saw in previous sections how to control switching by operating various engines. Here, using the master engine which flips one switch at a time, we extend this to include quantities. Exploration of the prior is straightforward. We generate ON/OFF transitions just as in Section 10.2.1. An ON-to-OFF transition is just that; the switch is turned OFF and the quantity ignored. An OFF-to-ON transition requires an accompanying quantity, which is best sampled anew with q re-sampled through eqns (10.10) and (10.9). That completes exploration of the prior.

The likelihood function will involve the quantities q, often enough as a Gaussian

$$\mathcal{L} = \exp(-\text{quadratic function of the } q\text{'s}) \qquad (10.11)$$

as from data linearly related to the quantities and subject to Gaussian noise. Spectroscopic data, for example, are very often of this form. Within nested sampling, the current likelihood constraint $\mathcal{L} > \mathcal{L}^*$ imposes a restriction on the permitted quantities. In particular, an individual q will be constrained within the interval between the roots q_- and q_+ of the quadratic equation $\log\mathcal{L} = \log\mathcal{L}^*$. If the quadratic equation has no real roots, or if the upper root q_+ is negative, that particular q cannot be present at all — and must already be OFF because the current state always obeys the constraint. Otherwise, the quantity in question (which has to be positive) is restricted to

$$q_0 < q < q_+ \qquad (10.12)$$

where $q_0 = \max(q_-, 0)$, which is equivalent through eqn (10.9) to a related range

$$u_0 < u < u_+ \qquad (10.13)$$

for its controlling variable u.

Suppose first that the range (q_0, q_+) exists and does *not* include $q = 0$ (i.e. $q_0 = q_- > 0$), so that the OFF state is prohibited. Necessarily, the switch must currently be ON, and a trial transition to OFF will certainly fail. Even so, we are still allowed to re-sample q within its range, by setting its controller as

$$u = \texttt{Uniform}(u_0, u_+) \tag{10.14}$$

So at least the 'rejected' transition is not wasted; it can re-equilibrate its quantity q.

Now suppose that the range (q_0, q_+) *does* include $q = 0$ (i.e. $q_0 = 0$), so that OFF is a valid state. An ON-to-OFF transition would always be accepted. An OFF-to-ON transition is accompanied by a new q, but this will only be accepted if it falls within the range, for which the probability is

$$\alpha = u_+ - u_0$$

Equivalently, compute α first, then with probability α flip the switch ON and re-sample q through eqns (10.14) and (10.9).

The only wasted computations here are the rejected OFF-to-ON transitions, which are only frequent if the prior favours ON (large π) while the likelihood favours OFF (small α, usually accompanying small q) — in practice a relatively unusual combination.

Other transition engines can be programmed similarly. For the transition of eqn (10.7) involving neighbouring switches, the change should be accompanied by an acceptance probability proportional to the destination's α

where $c = \max(\alpha_4, \alpha_5)$. This preserves detailed balance in the usual way, whilst minimizing the number of rejected transitions.

10.3.1 Programming the estimation of quantities in 'C'

As in Section 10.2.2, we allow five components, now coded as $M = 5$ quantities q, each of which has a prior probability, coded as $\texttt{PrON}$, of being ON. A component that is OFF has its quantity q set to 0, while a component that is ON has positive quantity with prior $\text{prob}(q) = 1/(1 + q)^2$, as in eqn (10.9) in units where $a = 1$. The likelihood function is Gaussian

$$\mathcal{L}(q) = \exp(-\chi^2/2)$$

with normalization factor ignored for simplicity, where χ^2 is the chi-squared misfit

$$\chi^2 = \sum_{k=0}^{N-1} (F_k - D_k)^2 ,$$

scaled to unit noise $\sigma_k = 1$, between the N actual data D and mock data F, namely

$$F_k = \sum_{j=0}^{M-1} T_{kj}\, q_j ,$$

the response matrix **T** of these simulated measurements being coded as Expt. The Object structure is unchanged, except that the binary switches s are up-graded to floating-point quantities q. The binary-tree scheme of transition rates is also unchanged. However, the logLhood function has to calculate the residuals resid $= F - D$, and thence χ^2, in order to reach the required log-likelihood value.

```
// apply.c    "QUANTIFICATION" NESTED SAMPLING APPLICATION
//            (GNU GENERAL PUBLIC LICENSE  software: © Sivia and Skilling 2006)

// Problem:       5 quantities q, which can be ON (+ve) or OFF (zero),
//                as measured by linear data with Gaussian noise.
// Inputs:
//    Prior(q)    is prob(ON) = PrON  with prob(q | ON) = 1/(1+q)^2
//    Likelihood  is exp(-chisquared/2)
//                   chisquared = SUM residual^2
//                   residual   = mock data - actual data
//                   mock data  = [Expt response].[quantities q]
// Outputs:
//    Evidence    is Z = INTEGRAL Likelihood(q) * Prior(q) dq
//    Posterior   is P(q) = Likelihood(q) * Prior(q) / Z
//    Information  is H = INTEGRAL P(q) log(P(q)/Prior(q)) dq
/*_____*/

#define n    100     // # Objects
#define MAX 1000     // # iterates
#define M    5       // # switches
#define Mplus 8      // power-of-2 ≥ M
#define N    3       // # data
#include "tree.c"    // binary tree procedures void PutRate, int GetRate
const double PrON[M] ={0.1, 0.2, 0.3, 0.2, 0.1}; // prior prob(ON)
const double Data[N] ={3,6,9};                    // actual data
const double Expt[N][M] ={{0,1,2,3,4},            // 1st data response
                          {0,0,3,2,1},            // 2nd data response
                          {3,2,1,2,3}};           // 3rd data response
/*_____*/

typedef struct
{
    double  q[M];             // quantities, 0 = OFF else prob(q)=1/(1+q)^2
    double  Tree[2*Mplus];    // binary rate tree
    double  logL;             // logLikelihood = ln prob(data | q)
    double  logWt;            // log(Weight), with SUM(Wt) = Evidence Z
} Object;
/*_____*/
```

```
double logLhood(          // logLikelihood function
double* q)                // quantities
{
    double resid;         // residual = (mock - actual) data
    int    j, k;          // component and data counters
    double C = 0.0;       // chisquared
    for( k = 0; k < N; k++ )
    {
        resid = -Data[k];
        for( j = 0; j < M; j++ )
            resid += Expt[k][j] * q[j];
        C += resid * resid;
    }
    return  -0.5 * C + sqrt(DBL_EPSILON) * UNIFORM;
}                         // jitter eliminates ties between likelihood values
/*_____*/
void Prior(               // Set Object according to prior
Object* Obj)              // Object being set
{
    double u;
    int    i;
// Initialize empty Tree of transition rates
    Obj->Tree[0] = Mplus;
    for( i = 1; i < 2*Mplus; i++ )
        Obj->Tree[i] = 0.0;
// Initialize object
    for( i = 0; i < M; i++ )
    {
        u = (UNIFORM < PrON[i]) ? UNIFORM : 0.0;
        Obj->q[i] = u / (1.0 - u);
        PutRate(Obj->Tree, i, (u > 0.0) ? 1.-PrON[i] : PrON[i]);
    }
    Obj->logL = logLhood(Obj->q);
}
/*_____*/
double TryQ(              // revised trial q[i]
double* q,                // quantities
int     i,                // id of quantity q[i] to be varied
double  logLstar)         // constraint
{
    double resid;         // residual = (mock - actual) data
    int    j, k;          // component and data counters
    double A, B, C, D;    // quadratic coeffs and discriminant
    double u;             // controlling variable for q[i]
    double min, max;      // range, of q[i] then of u
// "L > Lstar" is quadratic interval "A*x*x + 2*B*x + C ≤ 0.0"
    A = B = 0.0;     C = 2.0 * logLstar;   // minus chisquared on entry
    for( k = 0; k < N; k++ )
    {
        resid = -Data[k];
        for( j = 0; j < M; j++ )
```

```
                    resid += Expt[k][j] * q[j];
            A += Expt[k][i] * Expt[k][i];
            B += Expt[k][i] * resid;
            C += resid * resid;
        }
//  Find controlling interval
        if( A > 0.0 )                      //  q[i] does affect data
        {
        //  Solve quadratic for (qmin,qmax) relative to q[i]
            D = B * B - A * C;             //  discriminant
            if( D > 0.0 )                  //  distinct real roots
            {
                if( B > 0.0 )
                { min = -B - sqrt(D);  max = C / min;  min /= A; }
                else
                { max = -B + sqrt(D);  min = C / max;  max /= A; }
        //  Reset (qmin,qmax) relative to origin q=0
                min += q[i];    max += q[i];
        //  Restrict (qmin,qmax) to non-negative values
                if( max <= 0.0 )  max = min = 0.0;
                else if( min < 0.0 )    min = 0.0;
        //  Transform to controlling interval (umin,umax)
                min /= 1.0 + min;   max /= 1.0 + max;
            }
            else
                min = max = 0.0;       //  no real roots, so null interval
        }
        else
        {                                  //  q[i] unmeasured, so
            min = 0.0;   max = 1.0;    //  all u are in range
        }
//  Accept/Reject
        if( q[i] == 0.0 )                     //  entry state OFF
            u = (UNIFORM < max - min)         //  accept ON, sample u
               ? min + (max - min) * UNIFORM : 0.0;
        else                                  //  entry state ON
            u = (min > 0.0)                   //  reject OFF, re-sample u
               ? min + (max - min) * UNIFORM : 0.0;
        return  u / (1.0 - u);                //  trial quantity
}
/*_____*/
void Explore(             //  Evolve object within likelihood constraint
Object* Obj,              //  Object being evolved
double  logLstar)         //  Likelihood constraint L > Lstar
{
    double Interval = 30.0; //  pre-judged time to equilibrate
    double t = 0.0;         //  evolution time, initialized 0
    double qold;            //  entry value
    double logLtry;         //  trial logLikelihood
    int    i;               //  quantity being changed
    while( (t += -log(UNIFORM) / Obj->Tree[i]) < Interval )
    {
```

```
        i = GetRate(Obj->Tree);
        qold = Obj->q[i];      // enable recovery of entry state
        Obj->q[i] = TryQ(Obj->q, i, logLstar);   // trial state
        logLtry = logLhood(Obj->q);
        if( logLtry > logLstar )
        {
            Obj->logL = logLtry;                      // accept
            PutRate(Obj->Tree, i,
                    (Obj->q[i]>0.) ? 1.-PrON[i] : PrON[i]);
        }
        else
            Obj->q[i] = qold;                         // reject
    }
}
/*_____*/
void Results(        // Posterior properties, here statistics of s
Object* Samples,     // Objects defining posterior
int     nest,        // # Samples
double  logZ)        // Evidence (= total weight = SUM[Samples] Weight)
{
    double post[M] = {0,0,0,0,0}; // posterior prob(ON)
    double mean[M] = {0,0,0,0,0}; // quantity mean (when ON)
    double var[M]  = {0,0,0,0,0}; // variance (when ON)
    double w;                     // Proportional weight
    int    i, k;                  // Sequence and sample counters
    for( i = 0; i < nest; i++ )
    {
        w = exp(Samples[i].logWt - logZ);
        for( k = 0; k < M; k++ )
            if( Samples[i].q[k] > 0.0 )
            {
                post[k] += w;
                mean[k] += w * Samples[i].q[k];
                var[k]  += w * Samples[i].q[k]*Samples[i].q[k];
            }
    }
    for( k = 0; k < M; k++ )
    {
        mean[k] /= post[k];
        var[k] = var[k] / post[k] - mean[k] * mean[k];
        printf("%d: Posterior(ON) =%3.0f%%,", k, 100.*post[k]);
        printf(" q =%6.2f +-%6.2f\n", mean[k], sqrt(var[k]));
    }
}
```

As in the earlier pseudo-crystal module, procedure `Prior` sets the quantities according to their prior, before calculating the corresponding likelihood — including the
small jitter recommended for a discrete problem in which some transitions (specifically,
OFF-to-OFF) do not otherwise alter the likelihood. It also initializes the binary tree
of ON/OFF transition rates through `PutRate`. Procedure `Explore` is again slightly
up-graded. As well as flipping an ON/OFF state, it has to control the corresponding
quantity, which has to be stored (as `qold`) on entry in case the transition is rejected and

the entry state has to be recovered. Also, the task of proposing a new trial quantity is more complicated, so is delegated to a new procedure TryQ.

Procedure TryQ encodes the re-sampling suggested in Section 10.3 above. It starts by generating the residuals, from which the quadratic eqn (10.11) is derived, initially in terms of the deviation δq from the current value. Computation would be faster if the residuals were stored as part of their Object structure, and updated when necessary, but the code would be longer. The quadratic is solved to obtain the allowed q-range of eqn (10.12), which is transformed to the controlling u-range of eqn (10.13). Then u is appropriately re-sampled within its range, and the corresponding trial quantity returned. This proposed q is designed to obey the likelihood constraint automatically, but rounding errors or (more pertinently) likelihood-jitter may break the constraint. Whatever the cause, if the trial q results in an inadmissible likelihood, the trial is rejected by reverting to the original entry state.

Finally, the Results procedure accumulates the posterior probability of each quantity being ON (*What's there...?*), and the mean and standard deviation of that quantity if it's present at all (*...and how much?*). Brute force integration gives the true answers for comparison with the computed results.

	Correct	Computed
$\log_e Z$	−9.17	−9.64*
$\mathcal{H}$	7.59	7.63
prob(q_0 is ON)	99.6%, 2.28 ±0.39	99%, 2.20 ±0.43
prob(q_1 is ON)	11.6%, 0.61 ±0.68	17%, 0.63 ±0.71
prob(q_2 is ON)	99.6%, 1.76 ±0.32	99%, 1.76 ±0.37
prob(q_3 is ON)	9.6%, 0.43 ±0.42	11%, 0.54 ±0.49
prob(q_4 is ON)	2.4%, 0.19 ±0.16	4%, 0.14 ±0.15

* with ± 0.28 numerical uncertainty

The computed results are correct to within an unimportant random numerical deviation that can be further reduced by using more than 100 objects (and iterates) in the nested sampling.

10.4 Final remarks

In this book, we have tried to show that probability calculus is not just required for rational inference, but is also straightforward to use. What could be simpler than the sum and product rules? The algorithms with which we compute probabilistic results can be similarly straightforward, as we have attempted to demonstrate with short but powerful programs. Sadly, the research literature often seems a forbidding morass of extreme technicalities of doubtful importance presented in impenetrable jargon. Yet all this froth need not imply that important ideas are difficult. Indeed, our experience suggests the reverse, that the best ideas really are simple. We hope that in this book we have illuminated this simplicity sufficiently that Bayesian theory and practice become, like standard calculus, merely the automatic and un-regarded tools of our readers' professional activities.

A Gaussian integrals

The mathematical techniques and results used in this book should be familiar to most scientists and engineers from their first year undergraduate courses. If the Taylor series, partial differentiation, multiple integrals, vectors, matrices and so on only bring back hazy memories, then a browse through your favourite 'Mathematical Methods' text-book is strongly recommended; information on the (slightly) more advanced material used, such as Lagrange multipliers and Fourier transforms, is also likely to be contained therein. One topic which is often not dealt with in much detail, but is very useful in obtaining (at least approximate) analytical results in probability calculations, is that of Gaussian integrals; it is to this that we devote the next few pages.

A.1 The univariate case

Let's begin with the integral of a Gaussian function of just one variable

$$J = \int_a^b \exp\left[-\frac{x^2}{2\sigma^2}\right] \mathrm{d}x \,, \tag{A.1}$$

where we have taken the mean, or maximum, to be at the origin ($\mu = 0$) for simplicity. Despite its trivial look, this can only be evaluated easily when a and b are $\pm\infty$ or zero. In that instance, the non-obvious first step is to consider the square of eqn (A.1):

$$J^2 = \int_a^b \int_a^b \exp\left[-\frac{(x^2+y^2)}{2\sigma^2}\right] \mathrm{d}x \, \mathrm{d}y \,,$$

where we have multiplied through by an equivalent expression with y as the dummy variable (instead of x). If we now make the substitution $x = R\cos\theta$ and $y = R\sin\theta$, so that it's like changing from Cartesian to polars coordinates in Fig. 3.14, we can express J^2 as the product of two straightforward integrals:

$$J^2 = \int_0^{\theta_{\max}} \mathrm{d}\theta \int_0^{\infty} R\exp\left[-\frac{R^2}{2\sigma^2}\right] \mathrm{d}R \,,$$

where $\theta_{\max}$ is 2π if a and b are $\pm\infty$, and $\pi/2$ if they are 0 and ∞, respectively. Since the first term on the right reduces to just $\theta_{\max}$, and the second is equal to σ^2, we obtain

$$\int_{-\infty}^{+\infty} \exp\left[-\frac{x^2}{2\sigma^2}\right] \mathrm{d}x = \sigma\sqrt{2\pi} \,, \tag{A.2}$$

or half this value for the semi-infinite range. This is the origin of the normalization constant in eqn (2.14).

For situations where the limits are not $\pm\infty$ or zero, the integral of eqn (A.1) has to be evaluated numerically. We can either do the computation ourselves, perhaps by simply pressing a button on a calculator, or look up the answer in a standard table of results; the latter are listed in many books (e.g. Abramowitz and Stegun 1965) and often formally pertain to the *error function*:

$$\mathrm{erf}(z) \ = \ \frac{2}{\sqrt{\pi}} \int\limits_0^z \mathrm{e}^{-t^2}\,\mathrm{d}t\,, \tag{A.3}$$

which can be related to eqn (A.1) by

$$\int\limits_a^b \exp\left[-\frac{x^2}{2\sigma^2}\right]\mathrm{d}x \ = \ \sigma\sqrt{\frac{\pi}{2}}\left[\mathrm{erf}\left(\frac{b}{\sigma\sqrt{2}}\right) - \mathrm{erf}\left(\frac{a}{\sigma\sqrt{2}}\right)\right]. \tag{A.4}$$

If $a=-\sigma$ and $b=+\sigma$, then the integral is equal to 68% of the infinite case in eqn (A.2); for $\pm 2\sigma$, the fraction rises to just over 95%. Once the limits lie outside $\pm 3\sigma$ (from $x=\mu$), the result is virtually indistinguishable from $\sigma\sqrt{2\pi}$.

A.2 The bivariate extension

Having dealt with the one-variable case, let's consider a two-dimensional Gaussian. Again taking the maximum to be at the origin ($x_0 = y_0 = 0$), for simplicity, its most general form is

$$G(x,y) \ = \ \exp\left[-\tfrac{1}{2}\left(Ax^2 + By^2 + 2Cxy\right)\right], \tag{A.5}$$

where the three constants must satisfy the conditions $A>0$, $B>0$ and $AB>C^2$. First of all, let's integrate $G(x,y)$ with respect to just one of the variables; if we choose y, then the resulting marginal function $g(x)$ is

$$g(x) \ = \ \exp\left[-\tfrac{1}{2}Ax^2\right]\int\limits_{-\infty}^{+\infty}\exp\left[-\tfrac{1}{2}\left(By^2 + 2Cxy\right)\right]\mathrm{d}y\,, \tag{A.6}$$

where we will always take the limits to be $\pm\infty$ from now on, to facilitate an analytical solution. The integral on the right is most easily evaluated by rewriting the exponent as

$$By^2 + 2Cxy \ = \ B\left(y + \frac{Cx}{B}\right)^2 - \frac{C^2}{B}x^2\,,$$

a manipulation which is frequently called 'completing the square'; substituting this in eqn (A.6), we obtain

$$g(x) \ = \ \exp\left[-\tfrac{1}{2}\left(A - \frac{C^2}{B}\right)x^2\right]\int\limits_{-\infty}^{+\infty}\exp\left[-\tfrac{1}{2}B(y+\phi)^2\right]\mathrm{d}y\,,$$

where $\phi = Cx/B$. Since the y-integral is now like the one in eqn (A.2), apart from the unimportant offset ϕ, with $\sigma^2 = 1/B$, its value is equal to $\sqrt{2\pi/B}$; hence, the marginal function $g(x)$ is just a Gaussian centred at the origin:

$$g(x) = \int\limits_{-\infty}^{+\infty} G(x,y)\,\mathrm{d}y = \sqrt{\frac{2\pi}{B}}\,\exp\left[-\frac{x^2}{2\sigma_x^2}\right], \tag{A.7}$$

where the variance σ_x^2 is given by

$$\sigma_x^2 = \frac{B}{AB - C^2}. \tag{A.8}$$

By essentially repeating the procedure above, or by appealing to the symmetry of the problem, the corresponding integral of $G(x,y)$ with respect to x can also be shown to be a Gaussian centred at the origin:

$$\int\limits_{-\infty}^{+\infty} G(x,y)\,\mathrm{d}x = \sqrt{\frac{2\pi}{A}}\,\exp\left[-\frac{y^2}{2\sigma_y^2}\right], \tag{A.9}$$

where the variance σ_y^2 is given by

$$\sigma_y^2 = \frac{A}{AB - C^2}. \tag{A.10}$$

Finally, by integrating eqn (A.7) with respect to x, or eqn (A.9) with respect to y, using the result of eqn (A.2), we find that

$$\int\limits_{-\infty}^{+\infty}\int\limits_{-\infty}^{+\infty} G(x,y)\,\mathrm{d}x\,\mathrm{d}y = \frac{2\pi}{\sqrt{AB - C^2}}. \tag{A.11}$$

A.3   The multivariate generalization

To extend the preceding analysis to the case of a Gaussian function of many variables, we will have to draw quite heavily on some standard results from linear algebra. Let's begin by writing the multivariate generalization of eqn (A.5) in matrix–vector notation:

$$G(\boldsymbol{x}) = \exp\left[-\tfrac{1}{2}\,\boldsymbol{x}^{\mathrm{T}}\mathbf{H}\boldsymbol{x}\right], \tag{A.12}$$

where the transpose $\boldsymbol{x}^{\mathrm{T}} = (x_1, x_2, \ldots, x_N)$ and $\mathbf{H}$ is a (real) symmetric matrix, whose N eigenvalues $\{\lambda_j\}$ must all be positive. Thus eqn (A.5) represents the special case of $N = 2$, where the elements of $\mathbf{H}$ are given by: $H_{11} = A$, $H_{22} = B$ and $H_{12} = H_{21} = C$. To evaluate the N-dimensional integral of $G(\boldsymbol{x})$

$$Z = \iint \cdots \int G(\boldsymbol{x})\,\mathrm{d}x_1\,\mathrm{d}x_2 \cdots \mathrm{d}x_N, \tag{A.13}$$

let us rotate the axes by making the substitution

$$x = \mathbf{O}\,y\,, \qquad (A.14)$$

where the columns of the $\mathbf{O}$ matrix are the normalized eigenvectors of $\mathbf{H}$; since the latter are orthogonal to each other, this means that:

$$\left(\mathbf{O}^{\mathrm{T}}\mathbf{O}\right)_{ij} = \delta_{ij} \quad \text{and} \quad \left(\mathbf{O}^{\mathrm{T}}\mathbf{H}\mathbf{O}\right)_{ij} = \lambda_j\,\delta_{ij}\,, \qquad (A.15)$$

where δ_{ij} is equal to one if $i=j$, and zero otherwise. With this change of variables, the exponent in eqn (A.12) becomes

$$x^{\mathrm{T}}\mathbf{H}\,x = (\mathbf{O}\,y)^{\mathrm{T}}\mathbf{H}(\mathbf{O}\,y) = y^{\mathrm{T}}(\mathbf{O}^{\mathrm{T}}\mathbf{H}\mathbf{O})\,y = \sum_{j=1}^{N} \lambda_j\,y_j^2\,. \qquad (A.16)$$

Taking determinants, the first part of eqn (A.15) gives

$$1 = \det\left(\mathbf{O}^{\mathrm{T}}\mathbf{O}\right) = \det\left(\mathbf{O}^{\mathrm{T}}\right)\det\left(\mathbf{O}\right) = \left[\det\left(\mathbf{O}\right)\right]^2\,,$$

where the manipulations on the right follow from the rules that the determinant of a product of square matrices is equal to the product of their determinants, and that the swapping around of all the rows and columns of a matrix leaves the determinant unchanged; the Jacobian, $\left|\det\left(\mathbf{O}\right)\right|$, of an orthogonal transformation is therefore unity. Using this, the second part of eqn (A.15) gives

$$\lambda_1\lambda_2\ldots\lambda_N = \det\left(\mathbf{O}^{\mathrm{T}}\mathbf{H}\mathbf{O}\right) = \det\left(\mathbf{O}^{\mathrm{T}}\right)\det(\mathbf{H})\det(\mathbf{O}) = \det(\mathbf{H})\,. \qquad (A.17)$$

Accordingly, eqn (A.13) reduces to a simple product of one-dimensional integrals:

$$Z = \prod_{j=1}^{N} \int \exp\left[-\tfrac{1}{2}\,\lambda_j\,y_j^2\right]\mathrm{d}y_j\,.$$

Hence, using eqn (A.2), we find that

$$Z = \frac{(2\pi)^{N/2}}{\sqrt{\lambda_1\lambda_2\ldots\lambda_N}}\,. \qquad (A.18)$$

Finally, using eqn (A.17), the N-dimensional Gaussian integral of eqn (A.13) becomes

$$Z = \int \exp\left[-\tfrac{1}{2}\,x^{\mathrm{T}}\mathbf{H}\,x\right]\mathrm{d}^N x = \frac{(2\pi)^{N/2}}{\sqrt{\det(\mathbf{H})}}\,. \qquad (A.19)$$

As a simple check, this formula can easily be tested against the explicit $N=2$ result of eqns (A.5) and (A.11); there, $\det(\mathbf{H}) = H_{11}\,H_{22} - H_{12}\,H_{21} = A\,B - C^2$.

Apart from the normalization constant, or *partition function*, of eqn (A.18), the other quantity of interest is the covariance matrix σ^2. Its ijth element is formally defined by

$$\left(\sigma^2\right)_{ij} = \left\langle (x_i - x_{\mathrm{o}i})(x_j - x_{\mathrm{o}j}) \right\rangle\,, \qquad (A.20)$$

where $x_{oj} = \langle x_j \rangle$, and the expectation value of any function of the parameters $f(x)$ is given by the multiple integral

$$\langle f(\boldsymbol{x}) \rangle = \frac{1}{Z} \int f(\boldsymbol{x}) \exp\left[-\tfrac{1}{2}\boldsymbol{x}^{\mathrm{T}}\mathbf{H}\boldsymbol{x}\right] \mathrm{d}^N\boldsymbol{x} . \tag{A.21}$$

Since the maximum, or mean, of the multivariate Gaussian of eqn (A.12) is at the origin (so that $x_{oj} = 0$ for all j), eqn (A.19) becomes

$$\left(\sigma^2\right)_{ij} = \frac{1}{Z} \int x_i\, x_j \exp\left[-\tfrac{1}{2}\boldsymbol{x}^{\mathrm{T}}\mathbf{H}\boldsymbol{x}\right] \mathrm{d}^N\boldsymbol{x} . \tag{A.22}$$

By writing the exponent of the multivariate Gaussian in component form,

$$\boldsymbol{x}^{\mathrm{T}}\mathbf{H}\boldsymbol{x} = \sum_{l=1}^{N}\sum_{m=1}^{N} H_{lm}\, x_l\, x_m ,$$

we can see that the right-hand side of eqn (A.21) is related to the partial derivative of the logarithm of the partition function

$$-2\frac{\partial}{\partial H_{ij}}\left\{\log_e[Z]\right\} = \frac{1}{Z} \int x_i\, x_j \exp\left[-\tfrac{1}{2}\boldsymbol{x}^{\mathrm{T}}\mathbf{H}\boldsymbol{x}\right] \mathrm{d}^N\boldsymbol{x} .$$

Thus, in conjunction with eqns (A.18) and (A.21), we have

$$\left(\sigma^2\right)_{ij} = \frac{\partial}{\partial H_{ij}}\left\{\log_e[\det(\mathbf{H})]\right\} . \tag{A.23}$$

This strange-looking quantity can be evaluated by remembering that the determinant of a matrix is given by the *scalar product* of any row, or column, with its *cofactors*; this means that

$$\frac{\partial}{\partial H_{ij}}\left\{\det(\mathbf{H})\right\} = h_{ij} , \tag{A.24}$$

where h_{ij} is equal to $(-1)^{j-i}$ times the determinant of the $(N-1)$-squared matrix left by striking out the ith row and jth column of $\mathbf{H}$. Hence, eqn (A.22) becomes

$$\left(\sigma^2\right)_{ij} = \frac{h_{ij}}{\det(\mathbf{H})} .$$

As we are dealing with a symmetric matrix, h_{ij} is also the ijth cofactor of the transpose of $\mathbf{H}$; therefore, we finally obtain the result

$$\sigma^2 = \frac{\mathrm{adj}(\mathbf{H})}{\det(\mathbf{H})} = \mathbf{H}^{-1}, \tag{A.25}$$

where the *adjoint* of $\mathbf{H}$ in the numerator is a matrix consisting of the cofactors of $\mathbf{H}^{\mathrm{T}}$. This inverse relationship can again be easily checked for the $N=2$ case of eqn (A.5), by comparison with the results given in eqns (A.8) and (A.10).

B Cox's derivation of probability

Any general theory must apply to special cases. Following Cox (1946), we take this mantra to heart and consider merely the simple, unambiguous, 8-state toy world of up to three binary switches. We just want to be able to learn about it, and to distinguish plausible settings from implausible. Remarkably, consideration of this tiny world suffices to define the rules of probability calculus. Any other calculus leads to contradiction with how we wish to reason about it, and so no other calculus is acceptable to us. In his bibliography of work in the field, Jaynes (2003) remarks that in his view *'this article was the most important advance in the conceptual ... formulation of probability theory since Laplace'*. We concur, though we have re-ordered some of the material and modernized the approach. Cox wrote about general propositions. In modern idiom, we make this more specific by considering the binary switches with which we nowadays encode them.

Our belief about the state S of a system is always in a specific context X, and we write $\pi(S\,|\,X)$ for it. Thus, in the 1-bit context $I = \{\downarrow, \uparrow\}$ of a single switch A, we write our belief in A being '$\uparrow$' as

$$\pi(A\,|\,I), \quad \text{where} \quad (A\,|\,I) = \{\uparrow\} \mid \{\downarrow, \uparrow\},$$

and our belief in the converse 'NOT A' as

$$\pi(\overline{A}\,|\,I), \quad \text{where} \quad (\overline{A}\,|\,I) = \{\downarrow\} \mid \{\downarrow, \uparrow\}.$$

In the 2-bit context $J = \{\downarrow\downarrow, \downarrow\uparrow, \uparrow\downarrow, \uparrow\uparrow\}$ of two switches A, B, we have

$$(A\,|\,J) = \{\uparrow\downarrow, \uparrow\uparrow\} \mid \{\downarrow\downarrow, \downarrow\uparrow, \uparrow\downarrow, \uparrow\uparrow\}, \quad \text{with belief } \pi(A\,|\,J),$$

for the first bit A being '$\uparrow$', and similarly for the second bit B. There are also 'A AND B' joint beliefs about two bits both being '$\uparrow$',

$$(AB\,|\,J) = \{\uparrow\uparrow\} \mid \{\downarrow\downarrow, \downarrow\uparrow, \uparrow\downarrow, \uparrow\uparrow\}, \qquad \text{with belief } \pi(AB\,|\,J),$$

and other conditional assignments such as

$$(B\,|\,AJ) = \{\uparrow\uparrow\} \mid \{\uparrow\downarrow, \uparrow\uparrow\}, \qquad\qquad \text{with belief } \pi(B\,|\,AJ).$$

In the 3-bit context $K = \{\downarrow\downarrow\downarrow, \downarrow\downarrow\uparrow, \downarrow\uparrow\downarrow, \downarrow\uparrow\uparrow, \uparrow\downarrow\downarrow, \uparrow\downarrow\uparrow, \uparrow\uparrow\downarrow, \uparrow\uparrow\uparrow\}$ of A, B, C,

$$(A\,|\,K) = \{\uparrow\downarrow\downarrow, \uparrow\downarrow\uparrow, \uparrow\uparrow\downarrow, \uparrow\uparrow\uparrow\} \mid \{\downarrow\downarrow\downarrow, \downarrow\downarrow\uparrow, \downarrow\uparrow\downarrow, \downarrow\uparrow\uparrow, \uparrow\downarrow\downarrow, \uparrow\downarrow\uparrow, \uparrow\uparrow\downarrow, \uparrow\uparrow\uparrow\},$$
$$\text{with belief } \pi(A\,|\,K),$$

and similarly for other assignments.

We aim to develop a calculus for manipulating our beliefs about this system, and start by *asserting transitivity*—if, in context K, we have more belief in A than B, and more in B than C, then we assert that we have more belief in A than in C. To do otherwise would lead us to argue in circles. A consequence is that we can map π (whatever it was originally) onto real numbers, in which 'more belief in' is represented by '$>$'. The transitivity assertion is

$$\left. \begin{array}{l} \pi(A\,|\,K) > \pi(B\,|\,K) \\ \pi(B\,|\,K) > \pi(C\,|\,K) \end{array} \right\} \implies \pi(A\,|\,K) > \pi(C\,|\,K)\,.$$

So *beliefs are real numbers*—or at least they may as well be.

We now *assert that* knowing about A, and also about B given that knowledge, suffices to teach us about AB, all in the same overall context J. Some function F formalizes this inference:

$$\begin{array}{ll} (A\,|\,J) = \{\uparrow\downarrow, \uparrow\uparrow\} \,|\, \{\downarrow\downarrow, \downarrow\uparrow, \uparrow\downarrow, \uparrow\uparrow\}\,, & \text{belief } a = \pi(A\,|\,J)\,, \\ (B\,|\,AJ) = \{\uparrow\uparrow\} \,|\, \{\uparrow\downarrow, \uparrow\uparrow\}\,, & \text{belief } b = \pi(B\,|\,AJ)\,, \\ (AB\,|\,J) = \{\uparrow\uparrow\} \,|\, \{\downarrow\downarrow, \downarrow\uparrow, \uparrow\downarrow, \uparrow\uparrow\}\,, & \text{belief } \pi(AB\,|\,J) = F(a,b)\,. \end{array} \quad \text{(B.1)}$$

If F were independent of its second argument, while A was known to be '$\uparrow$', eqn (B.1) would say $\pi(AB\,|\,J) = F\big(\pi(\text{certainty})\big) = $ constant. Arbitrary beliefs about B would then all take the same value, which defeats our object. Likewise, if F were independent of its first argument, but B were known to be '$\uparrow$', all beliefs about A would take the same value. Hence, for a usable calculus, we need F to depend on both its arguments.

The world of three bits allows sequential learning too. The three beliefs

$$\begin{array}{ll} (A\,|\,K) = \{\uparrow\downarrow\downarrow, \uparrow\downarrow\uparrow, \uparrow\uparrow\downarrow, \uparrow\uparrow\uparrow\} \,|\, \{\downarrow\downarrow\downarrow, \ldots, \uparrow\uparrow\uparrow\}\,, & \text{belief } x = \pi(A\,|\,K)\,, \\ (B\,|\,AK) = \{\uparrow\uparrow\downarrow, \uparrow\uparrow\uparrow\} \,|\, \{\uparrow\downarrow\downarrow, \uparrow\downarrow\uparrow, \uparrow\uparrow\downarrow, \uparrow\uparrow\uparrow\}\,, & \text{belief } y = \pi(B\,|\,AK)\,, \\ (C\,|\,ABK) = \{\uparrow\uparrow\uparrow\} \,|\, \{\uparrow\uparrow\downarrow, \uparrow\uparrow\uparrow\}\,, & \text{belief } z = \pi(C\,|\,ABK)\,, \end{array}$$

chain together to define our belief $\pi(ABC\,|\,K)$. In the chain, B could be combined with A before linking with C, or with C before A. Hence

$$\pi(ABC\,|\,K) = F\big(\underset{AB|K}{F(x,y)},\ \underset{C|ABK}{z}\ \big) = F\big(\ \underset{A|K}{x}\ ,\ \underset{BC|AK}{F(y,z)}\,\big) \qquad \text{(B.2)}$$

in which the arguments of the outer F's are interpreted underneath. This is the 'associativity equation' and, following Lemma 1, it restricts F to be of the form

$$F(a,b) = w^{-1}\big(w(a) + w(b)\big)$$

where w is some invertible function of only one variable, instead of two. This is often quoted multiplicatively as $F(a,b) = w^{-1}\big(w(a) \times w(b)\big)$, but here we delay the final exponentiation of w and let the arguments add instead of multiply. Remembering that π was initially on an arbitrary scale, we can now upgrade to a less arbitrary scale of belief

$$\phi(\cdot) = w\big(\pi(\cdot)\big)$$

in which the sequential learning of eqn (B.1) proceeds by addition;

$$\phi(AB|J) = \phi(A|J) + \phi(B|AJ).\tag{B.3}$$

If we wish, we can revert to any other π in which learning proceeds by the appropriate modification of addition, but the new scale ϕ is available as a common standard. Yet there remains some arbitrariness, because ϕ could be rescaled by any constant factor whilst still obeying eqn (B.3). To fix the scale completely, we consider negation.

In the two-state context $I = \{\downarrow, \uparrow\}$ of a single bit, we *assert that* our belief about A defines our belief about its converse $\overline{A}$, formalized by some function f

$$\phi(\overline{A}|I) = f\big(\phi(A|I)\big).$$

Repeated negation is the identity, so

$$f\big(f(x)\big) = x.\tag{B.4}$$

Now consider the three-state context $T = \{\downarrow\uparrow, \uparrow\downarrow, \uparrow\uparrow\}$ of two bits A and B in which at least one is '$\uparrow$'. With context T understood throughout,

$$
\begin{aligned}
\phi(AB) &= \phi(A) + \phi(B|A) && \text{sequential learning}\\
&= \phi(A) + f\big(\phi(\overline{B}|A)\big) && \text{definition of } f\\
&= \phi(A) + f\big(\phi(\overline{B},\,A) - \phi(A)\big) && \text{sequential de-learning}\\
&= \phi(A) + f\big(\phi(\overline{B}) - \phi(A)\big) && B = \downarrow \text{ state is unique in } T\\
&= \phi(A) + f\big(f(\phi(B)) - \phi(A)\big) && \text{definition of } f\\
&= x + f\big(f(y) - x\big) && \text{name } \phi(A) = x,\ \phi(B) = y.
\end{aligned}
$$

Symmetry $AB = BA$ then gives

$$x + f\big(f(y) - x\big) = y + f\big(f(x) - y\big),\tag{B.5}$$

in which x and y are independent variables. Following Lemma 2, the functional eqns (B.4) and (B.5) together require

$$f(\xi) = \gamma^{-1}\log\big(1 - e^{\gamma\xi}\big),$$

where γ is a constant, and hence

$$\exp\big[\gamma\,\phi(\overline{A}|I)\big] = 1 - \exp\big[\gamma\,\phi(A|I)\big]\tag{B.6}$$

because $\xi = \phi(A|I)$ requires $f(\xi) = \phi(\overline{A}|I)$. We now upgrade to a new, fixed scale by defining $\mathrm{prob}(\cdot) = \exp\big[\gamma\,\phi(\cdot)\big]$. Qualitatively, $0 \leqslant \mathrm{prob}(\cdot) \leqslant 1$ because neither exponential in eqn (B.6) can be negative. Quantitatively, eqn (B.6) becomes

$$\mathrm{prob}(A|I) + \mathrm{prob}(\overline{A}|I) = 1 \tag{B.7}$$

which is the sum rule. Meanwhile, eqn (B.3) exponentiates to

$$\mathrm{prob}(AB|J) = \mathrm{prob}(A|J) \times \mathrm{prob}(B|AJ) \tag{B.8}$$

which is the product rule. We have derived the *sum and product rules* of probability calculus, and there's no scaling freedom left. With both rules obeyed, we are entitled to call $\mathrm{prob}(S|X)$ the *probability* of state S in context X. This quantification of belief is *what probability means*.

Though the calculus is now fixed, we are free to transform to other scales such as percentages $(100 \times \mathrm{prob})$, odds $\mathrm{prob}/(1-\mathrm{prob})$, logarithms $\log(\mathrm{prob})$ or whatever else might be convenient, provided that we transform the sum and product rules to compensate. For example, percentages add to 100, not 1. There is, of course, no change of content in any such reversible transformation.

For general inference, we use the simple switches $A, B, C, \dots$ to encode arbitrary propositions. Applying the product rule when we know B to be '↑', and hence that $\mathrm{prob}(A|J) = \mathrm{prob}(AB|J)$ because '↑↓' is excluded from our belief, shows that the true statement $(B|AJ)$ has unit probability: $\mathrm{prob}(B|AJ) = 1$. In general context, the unique true proposition thus has to be assigned $\mathrm{prob}(\texttt{TRUE}) = 1$. The negation of truth being falsity, it follows from the sum rule that $\mathrm{prob}(\texttt{FALSE}) = 0$. Hence

$$\mathrm{prob}(\texttt{FALSE}) = 0 \leqslant \mathrm{prob}(\cdot) \leqslant 1 = \mathrm{prob}(\texttt{TRUE}) \tag{B.9}$$

which identifies the range of probability values.

The Cox derivation rests only upon elementary logic applied to very small worlds. If there is a general theory of rational inference at all, it must apply in special cases, so it can only be this probability calculus. Moreover, any defined problem can be broken down into small steps. We have to use probability calculus in the small steps, and this implies using it overall in the larger problem. This is the only globally-applicable calculus we are ever going to have, so we should use it. And it does seem to be rather successful.

B.1 Lemma 1: associativity equation

Our task is to solve eqn (B.2), namely

$$F\big(F(x,y), z\big) = F\big(x, F(y,z)\big) . \tag{B.10}$$

Here and hereafter we assume that our functions are differentiable.

The first step is to find a special relationship between the derivatives of F. Defining

$$u = F(x,y) \quad \text{and} \quad v = F(y,z) , \tag{B.11}$$

and substituting in eqn (B.10),

$$F(u,z) = F(x,v) . \tag{B.12}$$

Taking the ratio $\partial_y \div \partial_x$ of differentials of eqn (B.12), with suffix r denoting derivative with respect to the r'th argument,

$$\frac{F_1(u, z)\, F_2(x, y)}{F_1(u, z)\, F_1(x, y)} = \frac{F_2(x, v)\, F_1(y, z)}{F_1(x, v)} . \tag{B.13}$$

Such relationships are non-singular because F is required to depend on both its arguments. Defining

$$g(\xi, \eta) = \log\big|F_2(\xi, \eta)\big| - \log\big|F_1(\xi, \eta)\big| \tag{B.14}$$

and using it in eqn (B.13),

$$g(x, y) = g(x, v) + \log\big|F_1(y, z)\big| . \tag{B.15}$$

Adding $g(y, z)$, and using eqn (B.14),

$$g(x, y) + g(y, z) = g(x, v) + \log\big|F_2(y, z)\big| . \tag{B.16}$$

Differentiating ∂_z(B.15),

$$0 = g_2(x, v)\, F_2(y, z) + \frac{F_{12}(y, z)}{F_1(y, z)} . \tag{B.17}$$

Differentiating ∂_y(B.16),

$$\frac{\partial}{\partial y}\Big[g(x, y) + g(y, z)\Big] = g_2(x, v)\, F_1(y, z) + \frac{F_{12}(y, z)}{F_2(y, z)} . \tag{B.18}$$

Eliminating F_{12} between eqns (B.17) and (B.18),

$$\frac{\partial}{\partial y}\Big[g(x, y) + g(y, z)\Big] = 0 . \tag{B.19}$$

The second step is to use this to express F in terms of functions of one variable only. Differentiating ∂_x(B.19),

$$\frac{\partial^2 g(x, y)}{\partial x\, \partial y} = 0 .$$

Solving in terms of arbitrary functions p and q of one variable,

$$g(x, y) = q(y) - p(x) . \tag{B.20}$$

Substituting eqn (B.20) in the exponential of eqn (B.14), using arbitrary functions P and Q related to exponentials of p and q,

$$\frac{Q(y)}{P(x)} = \frac{\partial F(x, y)/\partial y}{\partial F(x, y)/\partial x} . \tag{B.21}$$

Defining new functions R and S related to P and Q, and re-expressing F as Φ,

$$dR(x) = P(x)\,dx\,, \quad dS(y) = Q(y)\,dy\,, \quad F(x,y) = \Phi(R,S)\,. \tag{B.22}$$

Substituting from eqn (B.22) in eqn (B.21),

$$\frac{\partial \Phi(R,S)}{\partial R} = \frac{\partial \Phi(R,S)}{\partial S}\,.$$

Solving in terms of (the inverse of) an arbitrary function W,

$$\Phi(R,S) = W^{-1}(R+S)\,. \tag{B.23}$$

In terms of x and y,

$$F(x,y) = W^{-1}(R(x) + S(y))\,. \tag{B.24}$$

The third step is to remove excess generality in F, which appears to depend on as many as three functions of one variable. Substituting eqn (B.24) in eqn (B.12),

$$R(u) + S(z) = R(x) + S(v)\,. \tag{B.25}$$

Substituting eqn (B.24) in eqn (B.11),

$$W(u) = R(x) + S(y) \quad \text{and} \quad W(v) = R(y) + S(z)\,. \tag{B.26}$$

Eliminating $R(x)$ and $S(z)$ from eqns (B.25) and (B.26),

$$R(u) + W(v) - R(y) = W(u) - S(y) + S(v)\,. \tag{B.27}$$

Now u, y, v are independent variables derived from x, y, z, so the u- and v-dependences in eqn (B.27) require

$$R(\xi) = W(\xi) + c \quad \text{and} \quad S(\xi) = W(\xi) + d\,, \tag{B.28}$$

where c and d are constants. Substituting from eqn (B.28) in eqn (B.24),

$$W(F(x,y)) = W(x) + W(y) + c + d\,. \tag{B.29}$$

Defining

$$w(\xi) = W(\xi) + c + d$$

to eliminate $c + d$, and substituting in eqn (B.29),

$$F(x,y) = w^{-1}(w(x) + w(y))\,. \tag{B.30}$$

Finally, eqn (B.30) satisfies eqn (B.10) without further restriction, so it is the general solution.

B.2 Lemma 2: negation

We aim to solve for f obeying eqns (B.4) and (B.5), namely

$$f\big(f(x)\big) = x\,, \tag{B.31}$$

$$x + f\big(f(y) - x\big) = y + f\big(f(x) - y\big)\,. \tag{B.32}$$

The first step is to derive from eqn (B.32) a differential equation in only one variable. Defining

$$u = f(y) - x \quad \text{and} \quad v = f(x) - y\,,$$

and substituting in eqn (B.32),

$$x + f(u) = y + f(v)\,. \tag{B.33}$$

Forming the ratio $-\partial^2_{xy} \div (\partial_x \times \partial_y)$ of differentials of eqn (B.33) gives

$$\frac{f''(u)}{f'(u)\,[1 - f'(u)]} = \frac{f''(v)}{f'(v)\,[1 - f'(v)]} = \gamma\,, \tag{B.34}$$

because the separation of the u and v variables means that both expressions must equal a constant γ.

The second step is to solve this for f, disallowing the singular possibilities $f' = 0$ and $f' = 1$ because in each case eqn (B.32) would force the supposedly independent variables x and y to be equal.

$$
\begin{aligned}
f''(\xi)/f'(\xi) &= \gamma - \gamma f'(\xi) &&\text{re-arrange} \\
\log|f'(\xi)| &= \gamma\xi - \gamma f(\xi) + a &&\text{integrate} \\
f'(\xi) &= A \exp\big(\gamma\xi - \gamma f(\xi)\big) &&\text{exponentiate} \\
e^{\gamma f}\,df &= A\,e^{\gamma\xi}\,d\xi &&\text{separate} \\
e^{\gamma f} &= A\,e^{\gamma\xi} + B &&\text{integrate} \\
f(\xi) &= \gamma^{-1}\log\big(B + A e^{\gamma\xi}\big) &&\text{re-arrange} \tag{B.35}
\end{aligned}
$$

The third step is to remove excess generality in the solution. Substituting for f from eqn (B.35) in eqn (B.32),

$$\gamma^{-1}\log\big(A^2\,e^{\gamma y} + AB + B e^{\gamma x}\big) = \gamma^{-1}\log\big(A^2\,e^{\gamma x} + AB + B e^{\gamma y}\big)\,.$$

Since the x- and y-dependence both require $B = A^2$, eqn (B.35) becomes

$$f(\xi) = \gamma^{-1}\log\big(A^2 + A e^{\gamma\xi}\big)\,. \tag{B.36}$$

This is the most general solution of eqn (B.32).

The fourth step is to remove the remaining spurious generality by using eqn (B.31). Substituting eqn (B.36) in eqn (B.31),

$$\gamma^{-1} \log(A^2 + A^3 + A^2 e^{\gamma \xi}) = \xi.$$ (B.37)

Using the ξ-dependence in the exponential of eqn (B.37),

$$A^2 + A^3 = 0 \quad \text{and} \quad A^2 = 1.$$

Hence $A = -1$ so that eqn (B.36) becomes

$$f(\xi) = \gamma^{-1} \log(1 - e^{\gamma \xi}).$$ (B.38)

Finally, eqn (B.38) satisfies both eqns (B.31) and (B.32) without further restriction, so it is their general solution.

QED

Bibliography

An extensive list of references can be found in several advanced textbooks, such as:

Bernardo, J.M. and Smith, A.F.M. (1994). *Bayesian theory*. John Wiley, New York.

Jaynes, E.T. (2003). *Probability theory: the logic of science*. Cambridge University Press, Cambridge.

O' Hagan, A. (1994). *Kendall's advanced theory of statistics*, Vol. 2B: *Bayesian inference*. Edward Arnold, London.

Although still incomplete at his death, and edited to publication by G.L. Bretthorst, we would strongly recommend Jaynes' magnificent treatise to any serious student of the Bayesian approach; the tutorial paper by Loredo (1990) is also excellent. Two recent textbooks of note are MacKay (2003) and Gregory (2005), which are written from a scientific perspective similar to ours; by contrast, Bernardo and Smith (1994) and O' Hagan (1994), in common with many Bayesian texts, assume a more conventional statistical background.

Other general sources of relevant literature include the proceedings of the annual *Maximum Entropy and Bayesian Methods* workshops, published by Kluwer, and latterly the American Institute of Physics, and the four-yearly *Valencia International Meeting on Bayesian Statistics*, published by Oxford University Press. A list of the references specifically mentioned in the present text is given below.

Abramowitz, M. and Stegun, I.A. (1965). *Handbook of mathematical functions.* Dover, New York.

Acton, F.S. (1970). *Numerical methods that work.* Harper & Row, New York.

Aitken, M.J. (1998). *An Introduction to optical dating.* Clarendon Press, Oxford.

Allen, M.P. and Tildesley, D.J. (1987). *Computer simulation of liquids.* Oxford University Press, Oxford.

Bayes, T. (1763). An essay towards solving a problem in the doctrine of chances. *Phil. Trans. Roy. Soc.*, **53**, 330–418.

Bernardo, J.M. (1979). Reference posterior distributions for Bayesian inference. *J. Roy. Stat. Soc. B*, **41**, 113–47.

Bernoulli, J. (1713). *Ars conjectandi.* Thurnisiorum, Basel.

Bontekoe, T.R. (1993). Pyramid images. In *Maximum entropy and Bayesian methods* (ed. A. Mohammad–Djafari and G. Demoment). Kluwer, Dordrecht.

Box, G.E.P. and Tiao, G.C. (1968). A Bayesian approach to some outlier problems. *Biometrika*, **55**, 119–29.

Bretthorst, G.L. (1988). *Bayesian spectrum analysis and parameter estimation.* Lecture Notes in Statistics, Vol. 48, Springer–Verlag, Berlin.

Bretthorst, G.L. (1990). Bayesian analysis II: model selection. *J. Magn. Reson.*, **88**, 552–70.

Bretthorst, G.L. (1993). On the difference in means. In *Physics and probability* (ed. W.T. Grandy, Jr. and P.W. Milonni). Cambridge University Press, Cambridge.

Burg, J.P. (1967). Maximum entropy spectral analysis. In *Proceedings of the 37th Meeting of the Society of Exploration Geophysicists*, Oklahoma City. Reprinted 1978, in *Modern spectral analysis* (ed. D.G. Childers), IEEE press.

Bryan, R.K. (1990). Maximum entropy analysis of oversampled data problems. *Eur. Biophys. J.*, **18**, 165–74.

Caves, C.M., Fuchs, C.A. and Schack, R. (2002). Quantum probabilities as Bayesian probabilities. *Phys. Rev A*, **65**, 22305–10.

Charter, M.K. (1990). Drug absorption in man, and its measurement by MaxEnt. In *Maximum entropy and Bayesian methods* (ed. P.F. Fougère). Kluwer, Dordrecht.

Cox, R.T. (1946). Probability, frequency and reasonable expectation. *Am. J. Phys.*, **14**, 1–13.

David, W.I.F. (1990). Extending the power of powder diffraction for structure determination. *Nature*, **346**, 731–4.

David, W.I.F. and Sivia, D.S. (2001). Background estimation using a robust Bayesian analysis. *J. Appl. Cryst.*, **34**, 318–24.

de Moivre, A. (1733). Approximatio ad summam terminorium binomii $(a + b)^n$ in seriem expansi. Reproduced in Archibald, A.C. (1926). *Isis*, **8**, 671–83.

Fischer, R., Hanson, K.M., Dose, V. and von der Linden, W. (2000). Background estimation in experimental spectra. *Phys. Rev. E*, **61**, 1152–60.

Frieden, B.R. (1972). Restoring with maximum likelihood and maximum entropy. *J. Opt. Soc. Am.*, **62**, 511–18.

Gill, P.E. , Murray, M. and Wright, H.W. (1981). *Practical optimization*. Academic Press, London.

Gregory, P.C. (2005). *Bayesian Logical Data Analysis for the Physical Sciences*. Cambridge University Press, Cambridge.

Gull, S.F. (1988). Bayesian inductive inference and maximum entropy. In *Maximum entropy and Bayesian methods in science and engineering*, Vol. 1 (ed. G.J. Erickson and C.R. Smith). Kluwer, Dordrecht.

Gull, S.F. (1989a), Developments in maximum entropy data analysis. In *Maximum entropy and Bayesian methods* (ed. J. Skilling). Kluwer, Dordrecht.

Gull, S.F. (1989b). Bayesian data analysis: straight–line fitting. In *Maximum entropy and Bayesian methods* (ed. J. Skilling). Kluwer, Dordrecht.

Gull, S.F. and Daniell, G.J. (1978). Image reconstruction from incomplete and noisy data. *Nature*, **272**, 686–90.

Gull, S.F. and Skilling, J. (1984). Maximum entropy image reconstruction. *IEE Proc.*, **131**F, 646–59.

Jaynes, E.T. (1957). Information theory and statistical mechanics. *Phys. Rev.*, **106**, 620–30; **108**, 171–90.

Jaynes, E.T. (1963). Foundations of probability theory and statistical mechanics. In *Delaware seminar in foundations of physics* (ed. M. Bunge). Springer–Verlag, Berlin.

Jaynes, E.T. (1978). Where do we stand on maximum entropy? In *The maximum entropy formalism* (ed. R.D. Levine and M. Tribus). M.I.T. Press, Cambridge, MA.

Jaynes, E.T. (1983). *Papers on probability statistics and statistical physics* (ed. R.D. Rosenkrantz). Reidel, Dordrecht.

Jaynes, E.T. (1986). Bayesian methods: an introductory tutorial. In *Maximum entropy and Bayesian methods in applied statistics* (ed. J.H. Justice). Cambridge University Press, Cambridge.

Jaynes, E.T. (1989). Clearing up mysteries — the original goal. In *Maximum entropy and Bayesian methods* (ed. J. Skilling). Kluwer, Dordrecht.

Jeffreys, H. (1939). *Theory of probability*. Clarendon Press, Oxford.

Keynes, J.M. (1921). *A treatise on probability*. MacMillan, London.

Kirkpatrick, S., Gelatt, C.D. and Vecchi, M.P. (1983). Optimization by simulated annealing. *Science*, **220**, 671–80.

Knuth, K.H. (1999). A Bayesian approach to source separation. In *Proceedings of the first international workshop on independent component analysis and signal separation: ICA'99* (ed. J.-F. Cardoso, C. Jutten and P. Loubaton), Aussios, France, Jan. 1999, pp. 283–8.

Knuth, K.H. (2005). Informed source separation: a Bayesian tutorial. In *Proceedings of the 13th European signal processing conference (EUSIPCO 2005)* (ed. E. Kuruoglu), Antalya, Turkey.

Laplace, P.S. de (1812). *Théorie analytique des probabilités*. Courcier Imprimeur, Paris.

Laplace, P.S. de (1814). *Essai philosophique sur les probabilités*. Courcier Imprimeur, Paris.

Laue, E., Skilling, J. and Staunton, J. (1985). Maximum entropy reconstruction of spectra containing antiphase peaks. *J. Magn. Res.*, **63**, 418–24.

Loredo, T.J. (1990). From Laplace to supernova 1987A: Bayesian inference in astrophysics. In *Maximum entropy and Bayesian methods* (ed. P.F. Fougère). Kluwer, Dordrecht.

MacKay, D.J.C. (2003). *Information Theory, Inference, and Learning Algorithms*. Cambridge University Press, Cambridge

Metropolis, N., Rosenbluth, A.W., Rosenbluth, M.N., Teller, A.H. and Teller, E. (1953). Equation of state by fast computing machines. *J. Chem. Phys.*, **21**, 1087–92.

Michalewich, Z. (1992). *Genetic algorithms + data structures = evolution programs*. Springer–Verlag, New York.

Nelder, J.A. and Mead, R. (1965). A simplex method for function minimization. *Comput. J.*, **7**, 308–13.

Newton, T.J. (1985). Blind deconvolution and related topics. Ph.D. Thesis, Cambridge University.

Press, W.H., Flannery, B.P., Teukolsky, S.A. and Vetterling, W.T. (1986). *Numerical recipes: the art of scientific computing*. Cambridge University Press, Cambridge.

Refson, K. (2000). MOLDY: a portable molecular dynamics simulation program for serial and parallel computers. *Comput. Phys. Comm.*, **126**, 309–28.

Roberts, S. and Everson, R. (2001). *Independent component analysis: principles and practice*. Cambridge University Press, Cambridge.

Shannon, C.E. (1948). A mathematical theory of communication. *Bell Syst. Tech. J.*, **27**, 379–423 and 623–56.

Shore, J.E. and Johnson, R.W. (1980). Axiomatic derivation of the principle of maximum entropy and the principle of minimum cross-entropy. *IEEE Trans.*, **IT-26**, 26–37.

Sibisi, S. (1990). Quantified MaxEnt: an NMR application. In *Maximum entropy and Bayesian methods* (ed. P.F. Fougère). Kluwer, Dordrecht.

Sibisi, S. (1996). Compound Poisson priors. In *MaxEnt 96* (ed. M. Sears, V. Nedeljkovic, N.E. Pendock, S. Sibisi).

Sivia, D.S. and Carlile, C.J. (1992). Molecular spectroscopy and Bayesian spectral analysis — how many lines are there? *J. Chem. Phys.*, **96**, 170–8.

Sivia, D.S., Carlile, C.J., Howells, W.S. and König, S. (1992). Bayesian analysis of quasielastic neutron scattering data. *Physica B*, **182**, 341–8.

Sivia, D.S. and David, W.I.F. (1994). A Bayesian approach to extracting structure–factor amplitudes from powder diffraction data. *Acta Cryst. A*, **50**, 703–14.

Sivia, D.S. and Webster, J.R.P. (1998). The Bayesian approach to reflectivity data. *Physica B*, **248**, 327–37.

Skilling, J. (1988). The axioms of maximum entropy. In *Maximum entropy and Bayesian methods in science and engineering*, Vol. 1 (ed. G.J. Erickson and C.R. Smith). Kluwer, Dordrecht.

Skilling, J. (1989). Classic maximum entropy. In *Maximum entropy and Bayesian methods* (ed. J. Skilling). Kluwer, Dordrecht.

Skilling, J. (1990). Quantified maximum entropy. In *Maximum entropy and Bayesian methods* (ed. P.F. Fougère). Kluwer, Dordrecht.

Skilling, J. (1991). Fundamentals of MaxEnt in data analysis. In *Maximum entropy in action* (ed. B. Buck and V.A. Macaulay). Clarendon Press, Oxford.

Skilling, J. (1998). Massive inference and maximum entropy. In *Maximum entropy and Bayesian methods* (ed. R. Fischer, R. Preuss and U. von Toussaint). Kluwer, Dordrecht.

Skilling, J. (2004). Nested sampling. In *Maximum entropy and Bayesian methods in science and engineering* (ed. G. Erickson, J.T. Rychert, C.R. Smith). *AIP Conf. Proc.*, **735**, 395–405.

Skilling, J. and Bryan, R.K. (1984). Maximum entropy image reconstruction: general algorithm. *Mon. Not. R. Astron. Soc.*, **211**, 111–24.

Sonett, C.P. (1990). Atmospheric ^{14}C variations: a Bayesian prospect. In *Maximum entropy and Bayesian methods* (ed. P.F. Fougère). Kluwer, Dordrecht.

Stokes, S. (1999). Luminescence dating applications in geomorphological research. *Geomorphology*, **29**, 153–71.

Tikhonov, A.N. and Arsenin, V.Y. (1977). *Solutions of ill–posed problems*. Halsted Press, New York.

Twain, M. (1924). *Autobiography*, Vol. 1, p. 246. Harper and Brothers, New York.

Yethiraj, M., Robinson, R.A., Sivia, D.S., Lynn, J.W. and Mook, H.A. (1991). A neutron–scattering study of magnon energies and intensities in iron. *Phys. Rev. B*, **43**, 2565–74.

Index